» The Winter of Our
Disconnect

Author and social commentator Dr. Susan Maushart is a columnist for *The Weekend Australian Magazine* and an independent producer for ABC Radio Australia. Her previous four books have been published in eight languages. She holds a Ph.D. in media ecology from New York University.

» The Winter of Our

Disconnect

How three totally wired teenagers »

« (and a mother who slept with her iPhone)

pulled the plug on their technology »

and lived to tell the tale

SUSAN MAUSHART

PROFILE BOOKS

First published in Great Britain in 2011 by
PROFILE BOOKS LTD
3A Exmouth House
Pine Street
London EC1R 0JH
www.profilebooks.com

First published in the United States of America in 2011 by
Jeremy P. Tarcher/Penguin

Previously published in 2010 by Bantam Australia

10 9 8 7 6 5 4 3

Printed and bound in Great Britain by
Clays, Bungay, Suffolk

A CIP catalogue record for this book is available from the British Library.

ISBN 978 1 84668 464 7
eISBN 978 1 84765 740 4

The paper this book is printed on is certified by the © 1996 Forest Stewardship Council A.C. (FSC).
It is ancient-forest friendly. The printer holds FSC chain of custody SGS-COC-2061

» Contents

For Christine L. Nystrom
and
Neil Postman (1931–2003)

I would like to extend thanks to the Institute of Advanced Studies at the University of Western Australia, which provided a techno-haven for writing this book.

Raising three teenagers as a single parent is no Contiki cruise at the best of times. But when I decided we should all set sail for a six-month screen-free adventure, it suddenly came closer to *The Caine Mutiny*, with me in the Bogart role.

There were lots of reasons why we pulled the plug on our electronic media . . . or, I should say, why I did, because heaven knows my children would have sooner volunteered to go without food, water, or hair products. At ages fourteen, fifteen, and eighteen, my daughters and my son don't use media. They *inhabit* media. And they do so exactly as fish inhabit a pond. Gracefully. Unblinkingly. And utterly without consciousness or curiosity as to how they got there.

They don't remember a time before e-mail, or instant messaging, or Google. Even the media of their own childhood—VHS and dial-up, Nintendo 64 and "cordful" phones—they regard as relics, as quaint as inkwells. They collectively refer to civilization pre–high-definition flat screen as "the black and white days."

My kids—like yours, I'm guessing—are part of a generation that cut its teeth, literally and figuratively, on a keyboard, learning to say "'puter" along with "Mama," "juice," and "Now!" They're kids who've had cell phones and wireless Internet longer than they've had molars. Who multitask their schoolwork alongside five or six other electronic inputs, to the syncopated beat of the Instant Messenger pulsing insistently like some distant tribal tom-tom.

Wait a minute. Did I say they do their schoolwork like that? Correction. They do their *life* like that.

When my children laugh, they don't say "ha ha." They say "LOL." In fact, they conjugate it. ("LOL at this picture before I Photoshopped your nose, Mom!") They download movies and TV shows as casually as you or I might switch on the radio. And when I remind them piracy is a crime, they look at one another and go "LOL." ("Aargh, me hearty!" someone adds, as if to an imaginary parrot, and they LOL again, louder this time.) These are kids who shrug when they lose their iPods, with all five thousand tunes and Lord-knows-what in the way of video clips, feature films, and "TV" shows (like, who watches TV on a television anymore?). "There's plenty more where that came from," their attitude says. And the most infuriating thing of all? They're right. The digital content that powers their world, like matter itself, can never truly be destroyed. Like the Magic Pudding of Australian legend, it's a dessert bar that never runs out of cheesecake.

There's so much that's wonderful, and at the same time nauseating, about that.

The Winter of Our Disconnect—aka The Experiment (as we all eventually came to call it)—was in some ways an accident waiting to happen. Over a period of years, I watched and worried as our media began to function as a force field separating my children from what my son, only half ironically, called RL (Real Life). But to be honest, the teenagers weren't the only ones with dependency issues. Although a relatively recent arrival to the global village, I'd been known to abuse information too. (Sneaking my iPhone into the toilet? Did I have no self-respect?) As a journalist, it was easy to hide my habit, but deep down I knew I was hooked.

The Winter of Our Disconnect started out as a kind of purge. It ended up as so much more. Long story short: Our digital detox messed with our heads, our hearts, and our homework. It changed the way we ate and the way we slept, the way we "friended," fought, planned, and played. It altered the very taste and texture of our family life. Hell, it even altered the mouthfeel. In the end, our family's self-imposed exile from the Information Age changed our lives indelibly—and infinitely for the better. This book is our travel log, our apologia, our *Pilgrim's Progress* slash *Walden Pond* slash *Lonely Planet Guide to Google-free Living*.

At the simplest level, *The Winter of Our Disconnect* is the story of how one highly idiosyncratic family survived six months of wandering through the desert, digitally speaking, and the lessons we learned about ourselves and our technology along the way. At the same time, our story is a channel, if you'll excuse the expression, to a wider view—into the impact of new media on the lives of families, into the very heart of the meaning of home.

"Only connect," implored E. M. Forster in his acclaimed novel *Howards End*, published a century ago. It must have seemed like such a good idea at the time. In 1910, the global village was still farmland. The telephone had only recently outgrown the ridicule that first greeted it. The first commercial radio station was still a world war away. It had been a scant sixty years since the debut of the telegraph. ("What hath God wrought?" inventor Samuel F. B. Morse brooded morosely in the world's first text message.)

Ninety-nine years and one trillion Web pages later, "only connect" is a goal we have achieved with a vengeance. So much so that our biggest challenge today may be finding the moral courage to log off.

Today, some 93 percent of teenagers are online and 75 percent use cell phones, according to 2010 figures from the Pew Internet & American Life Project. Marketing data show 92 percent of teens own an iPod or MP3 player, while upward of two-thirds own their own computer (and access to one at home is near-universal). But the most provocative statistics are those that show how intensely our children interact with their media. In a large-scale study of young people who use media, conducted in 2005—ancient history already—up to a third told the Kaiser Family Foundation they were using multiple electronic devices simultaneously "most of the time." Researchers calculated that that meant the average American teenager was spending 8.5 hours a day in some form of mass-mediated interaction. And because media use in families is directly correlated with income, the figures were higher still in households at the more affluent end of the socioeconomic spectrum, and where parents were more highly educated.[1]

By 2010, when Kaiser updated the data, the media bubble continued to swell. Kids aged eight to eighteen had now increased their screen time by more than an hour and a quarter a day, from six hours, twenty-one minutes to seven hours, thirty-eight minutes—or the equivalent of an average working day, seven days a week. When multitasking was factored into the equation, the figure distended even further: to nearly eleven hours of what researchers now call "exposure." Add time spent texting and talking on cell phones—which the Kaiser folks did not even define as media—and the picture gets downright radioactive.[2]

For Generation M, as the Kaiser report dubbed these eight- to eighteen-year-olds, media use is not an activity—like exercise, or playing Monopoly, or bickering with your brother in the backseat. It's an environment: pervasive, invisible, shrink-wrapped around pretty much everything kids do and say and think. How adaptive an environment is the question—and the answer, not surprisingly, seems to depend entirely on whom you ask. The Pew Project found that, among teens, 88 percent are convinced that technology makes their lives easier. A decidedly more ambivalent 69 percent of parents say the same—although two-thirds also make some effort to regulate their children's use of media in some way (rules about safe sites, file sharing, time use, etc.).[3]

A 2007 Kaiser study found that nearly one in five parents believed there was no need to monitor their kids' screen time closely,[4] while the Pew research showed an astonishing 30 percent of parents believe that media have no effect on their children one way or the other.[5] Maybe that's wishful thinking. On the other hand, maybe it's not wishful enough. "One way or the other"—to me it's like saying the food we eat, or the air we breathe, or the communities we live in have no effect on us one way or the other. Or it could be these parents simply had a hard time imagining life outside the technological bubble—and, if so, who could blame them? Before undertaking this project, I had a hard time imagining it myself.

The question of whether communication technology makes our lives *easier* is much more specific, and less difficult to answer. Or is it?

When I first read the Pew research, it reminded me of a study I read in the *British Journal of Sociology* on the impact of domestic technology on housework.[6] Among other findings, the researchers discovered that having a washing machine and a dryer actually *increased* the time families—okay, women—spent on laundry tasks. For a start, people with dryers wash more clothes. And although a washing machine definitely makes it easier to get dirty clothes clean, it also raises the bar on the underlying question of "how clean is clean enough?" The new technology, in other words, solves an existing problem but in the process it creates a new and improved problem, and more laundry. It's a tale that the history of technological innovation tells us over and over again, as if on an endless loop of tape. The promise of "better living through technology!"—and you can take your pick which

one—is *always* a loaded deal, and often a paradoxical one as well. It
tends to be both true and untrue in equal proportions. Our technologies invariably start out as responses to a need. But over time, and in
subtle, unpredictable ways, they come to redefine that need.

So . . . how connected, I found myself wondering, is connected
enough? As a social scientist, journalist, and mother, I've always been
an enthusiastic user of information technology (and I'm awfully fond
of my dryer too). But I was also growing skeptical of the redemptive
power of media to improve our lives—let alone to make them "easier" or simplify them. Like many other parents, I'd noticed that the
more we seemed to communicate as individuals, the less we seemed
to cohere as a family. (Talk about a disconnect!)

There were contradictions on a broader scale too—and they
have been widely noted. That the more facts we have at our fingertips, the less we seem to know. That the "convenience" of messaging
media (e-mail, SMS, IM) consumes ever larger and more indigestible
chunks of our time and headspace. That as a culture we are practically *swimming* in entertainment, yet remain more depressed than
any people who have ever lived. Basically, I started considering a sce-
nario E. M. Forster never anticipated: the possibility that the more
we connect, the further we may drift, the more fragmented we may
become.

Or not. Because, just to complicate matters, I happen to believe
that the possibilities held out to us by media are hugely exciting. I am
not a Golden-Ager, lamenting the decline of the candle in a neon-lit
world. Not in the least. I love my gadgets (and I've got a gazillion
of 'em to prove it). I think my life is enhanced by technology. And I
know the world at large is. Yet the idea that there might be a media
equivalent of what micro-finance guru David Bussau calls "an eco-
nomics of enough" continued to occupy my thoughts.

It was an intriguing set of questions—and I was pretty sure I would *not* find the answers on Wikipedia. But how on earth could I test my hypotheses-slash-hunches?

That's when I remembered Barry Marshall—the Australian micro-biologist who won a Nobel Prize in 2005 for the simple but astounding discovery that stomach ulcers are caused by bacteria. Not stress, or spicy foods, or excess acid. Germs. Plain old germs. In retrospect, it seems so obvious. In the early eighties, Marshall's theory was dismissed as outlandish—especially by the pharmaceutical companies that underwrite the clinical trials by which medical research is tested. Frustrated but undaunted, Marshall decided to take matters into his own hands . . . indeed, into his own stomach lining. He swallowed some of the bacteria in question and waited to see whether he would develop an ulcer. He did. And the rest—give or take a decade of intensive further research—is history.

So it occurred to me: If Marshall could use his own life as a petri dish, why couldn't I?

(Gulp.)

Who We Are, and Why We Pressed "Pause"

I love technology
But not as much as you, you see.
But I still love technology,
Always and forever.
    —KIP'S WEDDING SONG, *Napoleon Dynamite* (2004)

When I first announced my intention to pull the plug on our family's entire armory of electronic weaponry—from the ittiest bittiest iPod Shuffle to my son's seriously souped-up gaming PC (the computing equivalent of a Dodge Ram)—my three kids didn't blink an eye. Looking back, I can understand why. They didn't hear me.

Well, they *are* teenagers. And they were busy. Uploading photos from last night's gathering, stalking a potential boyfriend's ex-girlfriend's Facebook friends, watching Odie the Talking Pug on YouTube ("I ruuuv ooooo," he howls to David Letterman). "Guys, are you listening?" I persisted.

"Can't you see we're doing homework, Mum?" my son replied irritably.

To be fair, it was the kind of thing I say a lot. Such as, "That's it—you're grounded for life!" or "Wait till your father gets home, young lady" (and I've been divorced for fourteen years). It probably sounded to them like just another in a long line of empty threats. It even sounded that way to me, to be honest. The urge to do a full-scale digital detox had been building for years. But it was more in the nature of a wistful but essentially ridiculous fantasy—like having a torrid affair with the Dalai Lama, or learning to tie a scarf four ways.

And then I reread *Walden*. (Note to self: Friends don't let friends reread Thoreau during an estrogen low.)

Walden—the story of the most famous stint in rehab in literary history—is my favorite book in the whole world, and I try to read it at least as often as I have a pap smear. I love *Walden* for lots of reasons, but mostly for its economy—the way it distills life and language to its most intoxicating essentials. You probably already know that it was written by transcendentalist Henry David Thoreau, who left his hometown of Concord, Massachusetts, in 1844 to conduct "an experiment in living" in the woods near Walden Pond. He lived there for two years in a wooden hut he built with his own hands, subsisting mainly on a monkish diet of wheaten cakes and pond fish. No neighbors. No running water. And, needless to say, no kids.

To be honest, I'd been thinking about running away to the woods myself a lot toward the end of 2008. It wasn't just the three teenagers I was wrangling: Anni, who'd just turned eighteen (terrifyingly, the legal drinking age in Western Australia, where we lived); Bill, fifteen, the man of the house (in his own mind, at least); and Sussy, the baby, fourteen ("Juliet's age when she got married, Mum," as she constantly reminded me).

They were at tricky ages, to be sure. But then, at age fifty, so was I. A career journalist, I was now part of the brand-new podcasting platform for ABC Radio. I loved the challenge of spitting out a weekly program, and I especially loved mastering the digital technology that modern broadcasting entails. What I didn't love was the huge time pressure. I was away from home more than I'd ever been since I'd started having babies, and the sense that I was losing control of my house and its contents—i.e., my kids—was ominous.

At the same time, our media habits had reached a scary kind of crescendo. It wasn't just the way the girls were becoming mere accessories of their own social-networking profile, as if real life were simply a dress rehearsal (or, more accurately, a photo op) for the next status update; or the fact that my son's domestic default mode was set to "illegal download," and his homework, which he'd insisted he needed a quad-core gaming computer and high-speed broadband to complete, was getting lost in transmission—although that was all part of it.

Thinking back, I realize there was no one breaking point, no single epiphany or *aha!* moment, but rather a series of such moments: scenes and stills I can scroll through in no particular order of importance, like a digital slideshow set to shuffle.

The abiding image of the back of Bill's head, for example, as he sat, enthroned before his PC in the region formerly known as the family room. Or the soundtrack of the conversations we'd been having for the last year or so, the ones that began with me saying anything at all ("Have you done your homework?" "Are you still enrolled in high school?" "Can you please put down your weapon and press 'pause' now? It's dinnertime") and ended with him replying, "Yeah. What?"

Maybe it was the evening the video clip playing on the corner of Sussy's desktop unexpectedly waved and called out gaily, "Hi, Susan!" It turned out to be a school friend streaming herself live on webcam via Skype. When my vital signs restabilized, I moved swiftly from simple fear to profound panic. What other visitors were logging on to her bedroom, in real time, full color, and stereo sound, while I slept?

Anni generally hit the trends first and most furiously. Always precocious, she'd been the first in her school to launch into MySpace way back in Year 10. (Not content with her own profile, she'd speedily created one for Jesus Christ [Relationship Status: It's Complicated] and another for Rupert, our pug [Favorite Movie: *Men in Black*].) At eighteen, she was still bingeing on social networking—Facebook being her drug of choice—and was also prone to sudden-onset gaming benders. Most recently, it was the online multiplayer word game called TextTwist. I'd watch her shoulders tense as she stabbed the keys with a viciousness normally reserved for conversations about curfews. And when she started gaining on her goal to become the world's number-one player, her jubilation had (for me) a disturbing edge. Watching her rapt, LCD-lit eyes, I couldn't help but think of Nero updating his status while Rome burned.

My own patterns were getting a little weird too. I never thought I'd be the kind of single mother who'd openly sleep with her iPhone, but . . . yeah. (I told myself it was no different from reading a book in bed—which, if I hadn't been watching feature-length movies and shopping for underwear, might well have been true.) In fact, if I didn't drag my laptop, a pair of speakers, my digital recorder, and a camera in too, I sometimes felt a little lonely. I told myself I was just doing my job. But there were times I looked less like a journalist than some demented IT technician in a nightie. Good times, good times.

However, it wasn't until I started surfing the Net, replying to text messages, listening to podcasts, and, on one memorable occasion, doing a live radio interview—all the while "otherwise engaged" in the loo—that I admitted I had a problem. Was I was using media to (cringe) self-medicate, on the fast track to becoming a middle-aged Lindsay Lohan of the App Store? Was it time to check myself in to rehab?

There was other stuff that was bothering me too. We were eating meals as a family less and less often. Never, if you want to get technical about it. The girls were either splurge-snacking or experimenting with weird diets. For days on end I swear Sussy ate nothing but condiments. Bill—aka the Cereal Killer—seemed to survive largely on shredded wheat and instant noodles, foods that shared a common, disturbing resemblance to roof insulation.

They were still having friends over, but more and more of their socializing took the form of little knots of spectators gathered around the cheery glow of YouTube—or, worse, dispersed into separate corners, each to his own device. Their sleep patterns were heading south too—hardly surprising given that the alerts from their three cell phones were intermittently audible through the night, chirping like a cadre of evil crickets.

And there were other things they'd hit the "pause" button on. Music—either playing it or listening to it as anything other than the background buzz to an instant messaging exchange. Books. Exercise. Conversation. And that other thing. Whaddaya call it? Oh, yeah. Life.

Although my own media habits were hardly immaculate, I could at least remember a time when things had been different. Simpler. More direct. Less tangled up with freaking USB cables. I found myself fantasizing about what life would be like in our house if I pulled the plug once and for all, hurtling us cold turkey into Wi-Fi withdrawal—myself and my omnipresent information IV included.

And at that stage, it *was* a fantasy. As a journalist and author, my livelihood depends on technology. People who wax nostalgic about a golden age of any kind, whether technological or political or cultural, have always seriously annoyed me. It's like listening to my mother talking about going to the movies for a quarter and having change left over to buy a hamburger and a Coke and, for all I know, stock options in MGM. The way I see it, it's hard enough to live in the present moment without somebody trying to drag you back to some sepia-tinged, hyperidealized pseudotopia that is usually three parts "La Vie en Rose" to one part irritable bowel syndrome. Every mythical "golden age," I have always believed, was exactly that. Mythical.

I grew up in the sixties and seventies, and although I have fond memories of *I Love Lucy*, instant mashed potatoes, and the Latin mass (in no particular order of importance), I do NOT believe my own childhood was superior to that of my own children. Parents and kids lived in two separate worlds in those days. That had its plusses, sure—like when you jumped on your bike and went to play at your friend's house till puberty, and nobody panicked. But it also had its minuses. Like most everybody else in my generation, I watched way too much dumb black-and-white TV, ate ridiculous snack food—come on, aerosol cheese?—and wouldn't have dreamed of confiding what I really felt and thought to a grown-up.

So nostalgia for "the way we were" isn't one of my weaknesses. I don't believe in avoiding your own reality, and I don't believe in the healing power of deprivation. The temptation to fix our family's discontents by ripping the modem from its socket smacked of both these fallacies.

Plus, I was menopausal. Sweet reason was not exactly what you'd call my strong suit.

If it hadn't been for Thoreau—or, more accurately, Sherman Paul, who wrote the introduction to my well-thumbed Riverside edition—I would probably have put away the idea with the rest of my hare-brained maternal schemes.* It was Paul's succinct explanation of why Thoreau took to the woods in the first place that was the tipping point. "He had reduced the means of life," Paul had written, "not because he wanted to prove he could go without them, or to disclaim their value in enriching life, but because they were usually factitious—they robbed one of life itself."

Thoreau's inspired mania for simplifying life, in other words, was just like Michelangelo's gift for "simplifying" a chunk of stone: "I saw

---

*A family uniform involving brown felt and Velcro; eating breakfast for dinner and dinner for breakfast; etc.

the angel in the marble and carved until I set him free." It was an act of creation and courage—not destruction, not fear. By isolating himself at Walden Pond, Thoreau hadn't run away from life. He'd run toward it. Why couldn't we leave our lives of quiet, digital desperation and do the same?

Now that I'd done the reframe—it wasn't something I'd be doing *to* my family, it was something I'd be doing *for* them!—I couldn't wait to begin. There was only one thing stopping me.

Oh, all right. Three things.

Anni, Bill, and Sussy, like most teenagers, live in a pre-Copernican universe. They are convinced the sun revolves around them. As their mother, I have done little to challenge this view. So when I finally worked up the courage to spring The Experiment on them for real, I chose my moment carefully. The stakeholders would need to be in a good mood. There would need to be lots of distractions: lights, music, refined sugar, whatever it took. And there would need to be witnesses.

Gracetown, Western Australia—go on, Google it—is a remote and ridiculously tiny coastal community on the southwest coast of Australia. It is renowned for its jaw-dropping Indian Ocean beaches, fearsome surf breaks, and curious lack of normal utilities. Gracetown is "electrified," but has no municipal water or gas supply—each house has its own rainwater tank and gas bottles—and no cell phone or Internet coverage. A persistent teenager climbing to the cliffside community's highest peak might get reception for a minute or two—and of course *all* teenagers are persistent—but that aside, we're talking Walden Pond with an Aussie accent.

So choosing to spend Christmas at Gracetown with our BFF (Best Friend Family) the Revells wasn't exactly a coincidence.

We'd arrived a few days early to settle into the rhythm. I'd insisted that everybody pack light, even the girls. It was one roll-on trunk of hair products each, I told them sternly, and no exceptions. And when Bill asked me if I'd seen his Nintendo DS, I thought, okay, this is my moment. I took a deep breath, looked him in the eye . . . and lied. I said I had no idea where it was. I had, in fact, hidden it at the bottom of a box of disused printer drivers the previous night, may the Good Lord have mercy upon my soul. "Let's read books in the car, honey," I suggested brightly. He muttered something under his breath. It was either "sick" or "suck," and I was pretty sure I knew which. In the end, I hauled out the iTrip, and we listened to podcasts on the three-hour car journey south—*This American Life* (my favorite radio show on any hemisphere), *The Hamish and Andy Show*, *The Moth*. Just being in a small space, listening to the same medium, made it feel like an old-fashioned family holiday already. But in a good way, in a good way.

Maybe I'd choked on the Nintendo thing, but the car trip had renewed my courage. Gracetown was the perfect setting in which to do the deed, even down to its name, with its faintly fundamentalist-slash-Elvis-impersonator overtones. It was just a case of choosing my moment. Christmas Day was only a few days hence, I reflected Grinchishly. Why not lower the boom then?

My dark thoughts about going off the grid dated as far back as Anni's four-year-old fixation with a certain *Lady Lovely Locks* video, featuring characters with names such as "ShiningGlory" and "Furball," and fiercely hair-driven plotlines. At the same time, like every other parent of toddlers, I was grateful for the thirty-minute break. (Bill's first sentence, which he bellowed solemnly at 5:15 every morning, was "Watch *White!*"—as in "The name's White. *Snow* White.") Yet as the years—and the technology—flew by, I rarely got beyond the grumbling stage. Occasionally I'd announce dramatically that I was "pulling the plug" so that everybody could read a book, or play a game, or just bicker with one another the old-fashioned way: face-to-face. Sometimes I'd even make good on it. "But I was doing my homework!" they'd wail, as the anime or the YouTube video or the MSN conversation froze mid-frame, exactly as if an evil fairy had waved her wand of doom. It felt good to pretend I still had some power in my own home. Deep down, though, even I—a woman so out of touch I still referred to "taping" shows on TV, as if they were packing boxes, or sprained ankles—was aware that ripping the modem out of the wall once every three or four weeks was a case of spitting into the Zeitgeist.

Who can ever say for certain what makes a person finally take that crucial leap into a life-changing decision? In my own case, I suspect The Experiment had roots as long and tangled as my fourteen-year-old's hair extensions. They probably went back to my graduate work in media ecology at New York University, my fascination with transcendentalist thinkers like Thoreau and Emerson, and my move to Australia in the late eighties.

There were more proximate causes too. One was an interview I did for one of my podcasts with a family of six kids, ranging in age from two to twelve, who were growing up entirely screen-free. Naturally I'd expected cult involvement, or at the very least a full-time parent-at-home. But no. Both parents worked as real-estate agents. There was no evidence of an extraterrestrial link. And the kids were amazing—full of excitement and ideas and trouvé collections and craft projects. Not fussy, adult-designed ones made from kits, but the kind you make from dead leaves and macaroni and toilet-paper rolls. They had a fort in the woods, and a tree swing, and a big dress-up box full of old clothes. "Don't you guys ever get bored?" I asked toward the end of the interview, almost desperate to find an edge to the story. But I already knew what the answer would be: a resounding "Nup." These kids knew they were a bit unusual, but they didn't feel deprived, if they thought about it at all—which, until the arrival of a woman with a microphone, I'm not sure they had. After all, their compensation for living without media was, to borrow Sherman Paul's phrase, nothing less than "life itself."

When I think it through, I realize there was all this backstory to my own decision. But reduce it to a sound bite and it was simply this: I was worried about my kids. About how they were using their time, and their space, and their minds. That's the center of gravity that pulled the whole thing together . . . and it's also, maybe, where my somewhat offbeat and bizarre life story crosses your own.

So, when I lowered the boom amid the happy detritus of a normal Australian Christmas morning—for chestnuts roasting on an open fire, substitute bacon and eggs on the barbie and the intoxicating whiff of 30+ sunscreen—there was nothing impulsive about it. Why I was making this decision was pretty clear in my mind. How I was going to obtain buy-in was a total blur. Granted, I do have kind of a gift for the pitch. In another, more lucrative life I would have made a bang-up used-car salesman. My enthusiasms—of which I have many—are as infectious as swine flu. My kids could tell you stories. Like the time I came home, flung open the door, and announced gleefully, "Hey, kids! Guess what?! I've lost my license for three whole months! Isn't it great? Because we are going to have *such* fun learning all about public transportation!" (It was just a few speeding fines. And not big ones either, until you added them all together. "Where I learned to drive—on the Long Island Expressway—anybody who *doesn't* go ten miles an hour over the speed limit is a pussy," I tried explaining to the constable. LOL he did not.)

I did such a consummate smoke-and-mirrors number when my marriage broke up that my eldest, who was four, literally didn't notice. "Where's Dada, anyway?" she finally inquired several weeks later. "Oh, didn't I tell you? He's got a cool new house and lucky you will get to stay there sometimes, just like Karen Brewer!" (Oh, for the days when a well-placed allusion to *The Baby-Sitters Club* was all it took to save one's sorry maternal ass!)

I don't lie, ever. Hardly. I sell. ("That's not a 'vegetable,' Bill. Why, that's a mouthwatering side dish of tender, buttery baby beans!") But let's face it: Spinning slightly overripe bananas to your toddlers is one thing—yes, I've been known to sing the Chiquita Banana song, and fake tap-dance too, if that's what it took. Selling your teenagers on the concept of giving up their information and entertainment lifelines for six months is quite another. To be honest, it kind of made giving birth in a manger in Bethlehem look like level-one Tetris.

Part of my strategy revolved around the presence of friendly witnesses: Mary and Grant and their teenage daughters, Ches and Torrie. Our fellow holiday-makers and oldest family friends would support me, and their presence would prevent any attempted wormouts. It was Mary who unexpectedly fed me my cue that morning, as she watched the girls unwrap their main presents—obscenely overpriced appliances hyped as "the Rolls-Royce of hair straighteners" (Lady Lovely Locks, may you rot in hell).

"But Suse," Mary blurted out. "Will they be able to use those when The Experiment starts?"

I shot her a look that could depilate, but it was too late. Everyone had heard her.

"Kids, I have an announcement to make," I began. All rustling of wrapping paper and gnawing of candy canes ceased. The girls put down their straighteners. Bill popped the lid back on his Sex Wax (a hair product, essentially, for surfboards). Even Rupert looked up with a mixture of anxiety and apprehension. But then he's a pug. He always looks like that. I took a deep breath and I hit them with it.

I didn't talk about being worried about their well-being, or their school performance, or their sleeping habits, or my fears for the arrested development of their social, intellectual, or spiritual skills. That would have been too much like nagging. It would have put them on the defensive. It would have started a conversation, and a conversation, frankly, was the last thing I wanted. The important thing was to announce, not to "suggest" or, heaven forbid, "discuss."

I concluded my announcement, eyes ablaze with missionary zeal (also fear), "It's an experiment in living. We are all going to do it together, as a family. And it's going to change our lives." There was a frozen pause. If life was a MacBook, this was our spinning color wheel of death.

Sussy broke the silence.

"You mean . . . like *Wife Swap*?" she asked.

"YES!" I roared. Bless the baby for throwing me a life raft. "Exactly like reality TV! Exactly! Except, of course, we won't have a TV . . ." I trailed off. I could see Bill and Anni exchange glances.

"What about homework?" Bill asked cannily.

"You can do it at the library, or at a friend's house, or at home using . . ."

"What? A stone tablet and a chisel?" Anni snapped.

"If you like," I replied evenly. (Pretending I don't get it is kind of my genius as a parent.) "But the point is, I can't control the universe. Alas. So it's only our *home* that's going to be screen-free."

I'd thought about this one a lot. In a perfect world—i.e., in which I did control the universe—The Experiment would be a *total* disconnect: no electronic media, at all, full stop, anywhere. It pained me to accept the reality that not even I could orchestrate such a thing. Short of moving to Djibouti, or imprisoning everybody in a backyard bomb shelter, there was no way I could pull it off. Like every other parent in the universe, I'd just have to find the serenity to accept the things I couldn't change, the courage to change the things I could, and sufficient download speed to tell the difference.

While they were still digesting the shred of good news I'd thrown at them, I added I'd be writing a book about our adventure. "Wait a minute, wait a minute," Anni interrupted. "A book? Like for money?"

"Maybe. Eventually," I allowed.

"Well, what do *we* get out of it?"

I winced. It was ugly, but I was ready. I knew that sooner or later we'd get around to talking turkey. As the eldest, and most practiced, plea-bargainer, Anni'd had plenty of experience in brokering damages claims on behalf of her plaintiffs. I could have quoted Thoreau. I could have explained the thing about Michelangelo, or produced a recommended reading list in media, cognition, and learning. Instead, Reader, I cash-incentivized them.

"Play along and play fair," I muttered, "and, yes, there'll be something in it for everybody." I sounded like a mafia boss. But *madre di dio*, I have three teenagers. What else am I supposed to sound like?

The proof of the pudding is in the eating, they say. And if there's one thing I've learned in my fourteen years of being a single parent, it's that a surprise attack—a pudding in the face, as it were—can be your best offensive strategy. I know that makes it sound as though you and your children are on opposing armies or something, but . . . well, aren't you? Boundary setting can be so hard, especially if, like me, you are secretly just a little intimidated by people who are more powerful, better looking, and wealthier than you are. Sure, they're your kids and you love them. But they can still be pretty scary.

That may be stating the case a little strongly. But as far as I can see, most parents of my generation—from the tail end of the Baby Boomers to the tender tip of Gen X—don't really rule the roost. We sort of scratch around it apologetically, seeking consensus.

We are bad at giving orders. But we are wonderful at giving options, and it's a habit that starts right from the git-go. "Milk, sweetheart?" we wheedle like some obsequious sommelier. "Our specials today are cow's, soy, breast or goat's." We ask our children to cooperate. We don't tell them to. And when there is an objection, we negotiate. I have one girlfriend who for many years paid her kids a weekly fee for brushing their teeth. I myself once slipped my seven-year-old a twenty for agreeing to a haircut. I think of that today and cringe. I'm sure I could have gotten it for ten.

So it's no wonder children today have a lively sense of entitlement. And that, metaphorically and otherwise, they take up more space. When I was growing up, back in "the black and white days," family life—and the distribution of family space—was very different. My sister and I shared a bedroom until we were teenagers, and so did most other kids we knew. We had a spare room, which my mother called the "sewing room." It housed a daybed that I never saw anybody use, and a Singer that dated from the Eisenhower administration. The last garment my mother had actually sewn was probably the petticoat for her poodle skirt, but that wasn't the point. The point was, unclaimed space belonged to the grown-ups, as if by divine right of mortgagee.

Today, most middle-class educated parents have reversed those priorities. Children are no longer the fringe dwellers of family life, but stake their claim to sit at the VIP table, even if they have to do it in a booster seat. They see resources in the home—furniture, appliances, food, adults, or any other standard utility—as *their* resources. I have a theory that this is particularly true in a single-parent home, where life tends to be more egalitarian, less structured and hierarchical. Okay, chaotic.

Let's consider the bedroom thing in more detail. Not only do our kids assume they are entitled to one, they also assume the right to maintain it according to a standard of their own choosing. My own kids were lisping, "But it's *my* bedwoom!" from the time they were able to toss Talking Elmo over the side of the crib. Today, every wall and surface of my son's private lair features sprayed-on graffiti—including, and this is no joke, a large corner cobweb—and his room is referred to affectionately as "the crack den." I've given my permission for all this largely because I'm insane, but also because I've internalized the mantra. It is, after all, his room.

By contrast, I didn't grow up with the sense that "my" bedroom was "mine" in the same way at all, and not just because it was shared. In the old days—and I'm talking waaay back, before parenting was even a gerund—"your" room belonged to your parents and everybody knew it. They controlled access. They chose the furniture and decorations. (My sister and I had white Queen Anne–style dressers, Martha Washington bedspreads, and a fake oil painting of Marie Antoinette—just to cover all the political bases. No wonder getting sent to our room was an effective punishment.) Parents also told you when and how to clean your room. We vacuumed and dusted and changed the bed linen every week. Once a month—and this messes with my mind even now—we *washed the woodwork*. When I told my own kids about this recently, their eyes grew as wide as laser discs. "Wow," Sussy mused. "What's woodwork?" Patiently, I pointed out the painted molding between the floor and wall in her bedroom. "Cool," she said politely. "I never noticed that before." I'd lay odds she'll never notice it again either.

Partly, I blame myself. When they were younger I had cleaners, like many other working parents. The beauty of that was not having to freak out so much about the housework. The tragedy was that it encouraged an "Elves and the Shoemaker" mentality: that cleaning and tidying were done magically in the dark of night by kind fairies. My own iron-clad habits of making my bed perfectly, complete with hospital corners and calculatedly "casual" pillow placement, are a source of genuine wonderment to my children. They used to bring in their friends sometimes for a peek, as if my bedroom were some exotic animal enclosure.

I know plenty of other parents who don't maintain *any* adult spaces—and who imply it's a form of fascism to try. Like the neighbors of mine whose double-story heritage home for many years bore a hand-painted wooden sign that announced CAMPBELL'S HOUSE, Campbell being their four-year-old son. (Campbell's mum and dad, you may not be surprised to learn, have since divided up Campbell's assets and gone their separate ways.) The way I see it, there are degrees of lunacy, and I like to think our own family inhabits a kind of happy medium between Campbell's House on one hand and *Brat Camp* on the other.

Kids don't just invade adult space, of course. They also invade adult time. And this is a commodity less easily cordoned off with bifold doors. Take "bedtime," for example. When I was a child it was nothing "to have to go to bed and see / the birds still hopping on the tree," as Robert Louis Stevenson somewhat sourly observed. Daylight savings or no daylight savings, you were tucked in at 6:30 p.m. And if you needed sunglasses and a UV blocker, so be it.

Today, "adult time" has become something that must be chiseled painfully from the bedrock of family life—or, more accurately, dug at its shoreline in haste, before the next high tide. I was strict about bedtimes when my children were little. (Sleep is to single mothers what helium is to a hot-air balloon.) But over the years, I bowed to the pressure to lighten up, from both their peers and my own. No woman wants to be seen as a control freak, least of all those of us who make our beds with a protractor and a spirit level.

Like so many other parents of my generation, I have grudgingly come to accept the prevailing view that "me time" is an indulgence— the temporal equivalent of a slab of mudcake or forbidden cigarette, the guilty exception to the rule that says to parents, and to mothers especially, and to single mothers most of all: "Your time is not your own." In the age of helicopter parenting, we are not supposed to want things any other way. Hovering, chopperlike, over our children's every move, as if they were escaped criminals or traffic accidents, is normal. In fact, we are supposed to be spinning our propellers with glee at the privilege. Being on call 24/7 is what having children is all about now—even children who are taller and more sexually active than you are (not that that's saying much in my case).

Opinions vary on whether the trend toward on-demand parenting is a healthy change. In a report on parenting college-age children, *New York Times* journalist Tamar Lewin spoke to one sophomore who uses more than three thousand cell phone minutes a month, most of it on calls to her parents, aunt, and grandparents. "I might call my dad and say, 'What's going on with the Kurds?' It's a lot easier than looking it up," she points out. Her father is good with this. "Whether you're wondering about a sweater or a class, it's great to have someone to bounce questions off. And why not a parent?" he asks genially.[1]

Um, maybe because the whole point of becoming an adult is to achieve self-reliance? Because maturity is largely about acquiring the confidence and the competence to make your own decisions? "Before the Industrial Revolution, there wasn't this concept that children should grow up, move away, and become autonomous," the father objects. That's very true. But this man's daughter doesn't live in an agrarian society. She lives in a dorm at Georgetown.

Whatever you think of its merits as a caregiving philosophy, there is no disputing that the helicopter parent is the bastard child of the Information Age. Without complex flight-control gear and a sophisticated communication network on constant alert, the level of surveillance we now regard as normal, even necessary, would be unthinkable. (More on that when we look in detail at the relationship messages we are sending with our cell phones—and I don't mean the SMSs—in Chapter Four.)

For now, suffice it to observe that children of all ages cross boundaries into adult territory like never before, and they do so because their parents have invited them to, whether consciously or not. I say that not in censure but in self-awareness. As a mother who once taught a graduate seminar while breast-feeding a five-month-old—and I mean literally *while* breast-feeding—I am a fully paid-up member of this parenting generation myself.

But more subversive than any of their incursions into adult time or space, I would argue, is our children's heightened sense of entitlement to *information*—promoted and protected by a Digital Bill of Rights under whose binding authority family life is being radically rewired.

"It's so unfair. I mean, what about their friends? Will they have any left at the end of it?"

"Surely they'll be bored."

"Forget about boredom. How will they do their schoolwork?"

"Those poor children!"

It's not often you get to eavesdrop on a conversation that's all about your own bad decision-making. In fact, barring my weekly phone call to my mother in North Carolina, I can't say I'd ever experienced it. I'm still not exactly sure how it happened. One minute I was saying good-bye to the father of one of Sussy's friends—we'd been confirming plans for an upcoming social event—and the next I was listening to a private conversation streaming live from his living room. After we'd hung up, his phone had somehow or other automatically redialed and . . . well, all I know is that I could hear Philip's voice plainly, only he wasn't talking to me. (Sometimes technology really *is* our friend.) Naturally, I responded the way any other intelligent, responsible adult would do in such a situation. I covered my right ear and jammed the phone to my left as hard as I could.

I could make out Philip explaining the outlines of our experiment to some unknown visitors, and something that sounded a lot like derisive snorting. No one actually came right out and said I was unhinged (although the phrase "a lot of pressure at work" got major air time). Nonetheless, the gist was clear: The Experiment was harsh and unworkable. The children would suffer. And I, as the crazed mastermind of it all, was borderline abusive, a cross between Supernanny and a guard at Abu Ghraib.

I wasn't surprised. I'd gotten this kind of reaction a lot since I'd started to "out" us. Even my agent Susan sounded a little worried when I first approached her with the idea for this book. "I love the idea," she wrote to me in an e-mail. "But are you sure you want to do this to your kids?" As if a child's right to Internet access and a cell phone plan were akin to her right to food and clothing and shelter and anti-frizz serum. Information starvation, the prevailing attitude suggested, was a form of child abuse—exactly as my kids had been trying to tell me all along!

There were others who cheered me on straightaway, including my stepdaughter, thirty-seven-year-old Naomi, who at that point was attempting to make a living renting virtual real estate on Second Life. But even among the yay-sayers, there was a widespread view that we were "going back to the seventies," or even going back to the nineteenth century.

What was *that* about?

In fact, most of the technologies that today rule our lives, just like the children that today rule our lives, emerged in the 1990s and early noughties. That means a lot of our media are teenagers too. Some of the most mesmerizing of them all—the iPod, the Nintendo Wii, Facebook—are barely toddlers. No wonder they're so damn attention-seeking.

Early on in The Experiment, we are in the car on the way to school when my fourteen-year-old reminds me to contact the school secretary, urgently, about my change of phone numbers. "I told her you didn't have a cell phone anymore and she got really mad," Sussy tells me. "She said, 'What if there's an emergency?'"

A stab of guilt goes through me like an overamped ringtone. "What did parents do ten years ago in an emergency?" I ask, feigning calm.

"Ten years ago," she replies coldly, "mothers stayed at home."

When I spoke to the school secretary later, she laughed. "I don't have a cell phone either," she admitted.

"Wow. Well, what do you do in an emergency?" I couldn't resist asking.

"What everybody used to do ten years ago. If there's a real emergency, don't worry, we'll find you."

The assumption that uninterrupted access to electronic information and entertainment is every child's right—and every parent's responsibility—has taken hold at a very deep level. Yet it has happened in the proverbial heartbeat.

Our own family's first dial-up account went online in 1996—and by Australian standards we were early adopters. Today, only 25 percent of American households do *not* have Internet access at home, according to 2010 figures from Nielsen NetRatings. In 1996, families were just discovering e-mail, and the more adventurous kids were test-driving the new search engines and stumbling upon free game sites.

Today, the average American child spends almost as much time online as he or she does *sleeping*.

Although the first IBM PC hit the market in 1981—think "Bette Davis Eyes" and Charles and Diana's wedding—most families didn't buy their first home computers until the late eighties. They were massive clunky things, with spooky green screens and less memory than an advanced Alzheimer's victim. The first "kids' computer" we acquired ran on Windows 98, and, as every grown-up knows, 1998 was, like, five minutes ago. When my eight-year-old daughter's friend customized the desktop and font colors for us, we stood around, wonder-struck. "She's a genius!" we agreed.

I bought our first cell phone in 2001—for my then-ten-year-old daughter, Anni, whose classmates gathered round in awe, begging for a demonstration. Today, the average high school kid spends an hour and a half text messaging each day, according to the 2010 Kaiser study. I didn't get my own phone for a year or so, and when I sent my first text message, I considered it a technological triumph second only to opening an e-mail attachment. iTunes was launched in 2001 too—although it took our family five long years to discover it—and the BlackBerry smartphone, albeit featuring a pretty dumb monochrome display, followed in 2002. By 2008, when the iPhone made its Australian debut, I had evolved from mild technophobe to a fully fledged geek. I had one within twenty-four hours.

Remember Game Boy? It may not have changed the world, but it sure revolutionized the family fly-drive vacation. It was first released in 1989. Nintendo 64, somewhat confusingly, came out in 1996, ten years before the release of the Wii. GameCube, Xbox, PlayStation, and their multitudinous handheld spawn—along with the other big names that have given joystick to the world—are children of the present millennium too.

The first MMOs—"massively multiplayer online" games such as World of Warcraft or Second Life, which generally involve simulation and role-playing—started appearing around the time my stretch marks did, in the early nineties. High-speed Internet—the fast and furious kind that has made it possible to *live* in cyberspace—has been available to domestic users for little more than a decade, and much less than that in Australia. In September 2008, the number of broadband subscribers in Australia was 5.7 million, having grown by 90 percent over the previous six months.[2] In the United States, despite an economy barely edging toward recovery, some 73 million American households, or about 60 percent of the total, held broadband subscriptions in 2010.[3]

TiVo (2000-ish) and even DVDs (1995-ish)—initials, its founders insist, that stand for nothing at all—seem like old technology already, though they're far from obsolete. But tracking down a VCR to watch your old home movies is like trying to find a Bakelite phone at the Apple Store.

Okay, what about e-mail? We can't talk indispensable, omnipresent, and omnivorous without talking e-mail.

I remember vividly the first e-mail message I ever received. It was from my girlfriend Pat who worked at Princeton and it was hand-delivered to me as hard copy (ironic, I know) by the IT technician at the university department where I was then employed. The year was 1994. Up until that day, I had only the fuzziest idea what e-mail *was*; I certainly didn't know I'd been allocated an account. I was thrilled. Deeply confused, but thrilled.

In fact, although e-mail was first demonstrated at MIT in 1961, it wasn't really until the late nineties that the ranks of business- and then home-users began banking up—particularly after the launch of Hotmail in 1996. And I know all this thanks to Wikipedia, of course (formally launched in 2001, in case you're offline and want to know).

Speaking of monoliths, Internet search site (and so much more) Google, which as I write employs a full-time global workforce of 20,222 and is regarded as the most powerful brand in the world, was registered as a domain in 1997. The ubiquitous verb "Google" was added to both the *Merriam-Webster's* and the *Oxford English Dictionary* in 2006. (For 1,490,000 other sources of information on this topic, just Google "history of Google.")

As if these statistics aren't startling enough, consider that Facebook, which as of this writing has 400 million active users—half of whom, I swear, are "friends" with my eighteen-year-old—was launched in 2004. *2004, people!* If Facebook were your child, it would still be in first grade. Today, Australians spend nearly one-third of their entire online time budget Facebooking. Admittedly, that's a world record. (Something to do with the ozone layer?) The rest of the global village isn't far behind. A Nielsen study published in January 2010 showed social networking increased internationally by 82 percent over the previous year.[4] A GFC (global friending crisis) appears unlikely anytime soon.

So new, and yet so far, eh? And if 1996 seems like the Stone Age even to someone like me—who can remember when the transistor frigging radio was cutting edge—is it any wonder that our kids can't imagine life without media? Or that on Maslow's Hierarchy of Needs (teenage edition), access to Internet browsing, an e-mail account, Facebook, iTunes, Nintendo, and a cell phone sits somewhere between "Safety" and "Love/Belonging"?

And yet, if the Digital Bill of Rights increasingly governs family life, and I would argue that it does, it's important to recognize who ratified it in the first place. Actually . . . and this is kind of embarrassing . . . we did. Especially those of us for whom the Information Age has coincided with our coming of age as parents, producing excitement, confusion, and a weird eclipse of attention. We've been caught up in a monsoon of technological change as mind-blowing in its intensity as . . . well, having kids in the first place. And I say that not to inspire guilt—if you're parents, you don't need my help with that one—but to raise consciousness.

Public debate around the media ecology of family life has had a helpless quality, positioning parents, not entirely inaccurately, as the little Dutch boy with his finger in the digital dyke. After six months of trying to keep a single household screen-free, trust me, I have much sympathy with that view. It is true, we don't have a prayer of holding back the flood entirely, even if we wanted to. But why would we want to? Information, like water, is a good thing . . . in its place.

The old saw reminds us that, to a man with a hammer, the whole world looks like a nail. Does it follow that to a girl with a Photo-bucket account, the whole world looks like a fashion shoot? Or that to a boy with a joystick and a graphics card, the whole world looks like a psychotic dwarf with an ax? To an important extent—definitely more than we have been comfortable admitting—yes, it sort of does. Ulti-mately, the answer is not to take away the hammer, but to see that it is used for more than bashing away at things. To ensure our children free their hands—and their heads—to take up other tools too.

We don't know who discovered water, but it wasn't a fish, some-one wise once observed. Whatever else it might accomplish, or fail to, our Experiment was about to propel us, stunned and gasping, out of our fishbowl for good.

Power Trip: The Darkness Descends

Only that day dawns to which we are awake.
—WALDEN, *chapter 18*

Slipping through the French doors onto the verandah, I felt the night air on my skin like some expensive moisturizer, warm and thick and lightly fragrant. It was just after midnight on a sultry summer night. I could see the moon and a few streaks of fast-moving cloud framed between the sloping tin roof and the white bougainvillea that had grown with fairy-tale abandon since the hot weather set in. The whiff of ocean salt was so fresh I could taste it.

Inside the house, the kids were asleep, sunburned and still sandy from two weeks of holiday. Concentrating, I could hear the drone of a ceiling fan and, farther back, the hum from my son's PC, as familiar and insistent as my own pulse. The digital display from my clock radio, flaring red, was just visible from where I stood. Seeing it flash, my heart began to beat in my head like a boom box.

I knew what I had to do. I was scared shitless to do it.

A voice within me spoke. "You're a parent, right? So what else is new?"

I looked squarely at the meter box mounted at eye level in front of me, the switch marked "MAINS" illuminated in a cheesy shaft of moonlight, took aim with a steady hand—and fired.

The idea to go screen-free for six months had been a calculated one. The idea to get in shape for it with two weeks of Blackout Bootcamp was more of a sudden inspiration. If you will, a lightbulb going off.

Psychologically, pulling the plug on the whole catastrophe—lights and appliances included—seemed to make sense. Like jumping into a cold pool, it was better, surely, to take the plunge in one breath-defying leap than to experience withdrawal gradually, degree by painful degree. And there was a bonus: By the time we got around to switching on the power again, we'd be desensitized. We'd rejoice in what we'd recovered, rather than bemoan what we'd lost. Or so, at least, I prayed.

In the meantime, well, I was quite fond of candlelight. (What woman of a certain age is not?) Plus, it would mean no vacuuming for two weeks, and no gigantic loads of washing. I told the kids we would each be responsible for our own laundry, and I could practically see them silently counting their pairs of clean underpants. We had a gas stove, so cooking wasn't an issue. And our gas water heater meant we could still have hot showers.

"It'll be like camping, guys!" I enthused.

"We hate camping," Sussy pointed out. "*You* hate camping, Mum."

Details, details! "I hate bugs and dirt and sleeping bags," I corrected. "This is camping the way it ought to be: with our own beds, pillows, stemware, and dual-flush toilets."

Thoreau himself helped me figure out the timing. He'd begun his life in the woods at Walden Pond on Independence Day, July 4, 1844. For us in the southern hemisphere—where Thoreau the naturalist would have been enchanted to discover the trees shed their bark but keep their leaves—that equated to January 4. Equally important, it would allow exactly twelve hours upon our return from Gracetown to machine-wash the station wagon full of holiday laundry we'd carted home with us. (Lord knows, doing it by hand would have been like vacuuming the house with an ear syringe.) Fittingly, from the children's point of view, our experiment would *end* on Independence Day.

"But what about the phone?" the kids had asked, panicked. I explained we'd simply go back to using old-fashioned cord phones, but they were still suspicious. "How will they stay charged?" they fretted.

"You can't be serious," I replied.

The truth is, I had no idea. Seriously, why *doesn't* a basic phone need electricity? It's kind of magical, once you stop to think about it. And actually I had stopped to think about it not that long ago.

It was on a day maybe six months earlier, when The Experiment was just an evil gleam in Mummy's gimlet eye. I was working from home, and I heard the sound of many phones ringing. Nothing unusual there. Our cordless collection at that point numbered five handsets, each programmed by Sussy with its own very special, faintly satanic ringtone (a techno-inspired remix of "Home Sweet Home" being the most chilling). Visitors occasionally cried out in terror when our phones rang. For me, the scary part was finding the damn things.

The children rarely bothered returning the phones to their recharging stations. To be honest, it wasn't my forte either. Most of the time, we'd simply drop them where they'd last been used, like gum wrappers or gym socks. Usually you could follow a ringtone to its source—tangled up in the bedclothes, or peeking out coyly from a drawer, or squashed under the sofa cushions like a raisin. But all this took time and energy and the kind of playful ingenuity I rarely had anymore, unless a pitcher of margaritas was involved. Even worse was the problem of the MIAs: phones that had wandered away from base and been left to die in some foxhole. Every once in a while they would *all* go missing, and I'd find myself on a grim telephonic scavenger hunt, seeking wounded handsets too weak to respond to signal.

On this particular day, I was expecting an important work call, and . . . well, let's put it this way. If a phone rings in the forest, and there's nobody there who can find it, does it still make a sound?

The next day I went out and bought a couple of old-fashioned plug-in phones for my bedroom and home office. The color of prosthetic limbs, they featured oversized buttons and pretty much nothing else, and were clearly meant for the demented and the infirm. So be it. At this point, I qualified easily on both counts. The kids ROFLed uproariously at the sight, but I knew I would have the last LOL. Next time the phone rang, instead of having to smoke it out with a gasoline torch, a psychic, and a Geiger counter, I could just pick it up and answer it. "Honestly!" I crowed. "What *will* they think of next?"

So we were good for phones. Illumination would be a snap too, once we'd stocked up on candles, kerosene, flashlights, and battery-operated lanterns. It would be like mood lighting. At the very least, it would cast a romantic glow on the buildup of unvacuumed pet hair. Food and drink might be a bit trickier, but I figured it would be a good excuse to buy a massive ice chest—I'd always wanted one with wheels, the kind that looks big enough to store human remains—and to shop for meals like a single person, i.e., spontaneously and one day at a time.

"Let us not be upset and overwhelmed in that terrible rapid and whirlpool called a *dinner*," Thoreau exhorted. I was down with that.

Coping without air-conditioning in the blast-furnace heat of the Western Australian summer was a less sanguine prospect. This, I knew, we would definitely yearn for . . . exactly as we did every other year. It's true. Despite temperatures that stay in the nineties all summer and frequently soar above 100, most West Aussies still live without climate control, and we were among them. It wasn't so bad in the port city of Fremantle, where we live, and where, even on the hottest days, the Fremantle Doctor—the famous sea breeze off the Indian

Ocean—comes to the rescue by early afternoon. Giving up our ceiling fans would be tough though—especially for Sussy, who prefers to sleep in a stiff breeze in all weathers, like a wolf cub.

If worse came to worst, I reflected, we could always fill up the big clawfoot tub with cold water and soak ourselves like navy beans. We used to do this when the kids were littler and more biddable—and when it gets really hot and no one's looking, I still do. It's not the most dignified way to chill out, but once you break through the pain barrier and cross over to feeling like human luncheon meat, it's way cool. Bill talked about reviving his favorite childhood cool-down strategy and making an "ice baby": a dishtowel packed with ice cubes and fastened with a rubber band, which one takes to bed and hugs like a new teddy bear, or a transitional husband.

The only thing I really worried about was my hair (a topic on which Thoreau provided not the slightest scrap of inspiration, incidentally). Although I'd tried to downplay it all their lives, the truth is, Anni and Sussy had a genetic predisposition to hair-related OCD. I had codependency issues with my straightener too—especially since I'd stopped coloring my hair. Anne Kreamer's book *Going Gray* had been one of the highlights of my literary year, and I'd become a complete convert to the cause. So far, the experience had been reasonably positive. I looked less heritage-listed than I'd feared—more Susan Sontag than Bob Hawke. I'd recently had bangs cut and found that with daily straightening, it looked borderline chic. Without daily straightening, alas, it looked borderline freak: wavy and cowlicked, like the warden of a women's prison on the late movie.

It's funny the things you cling to when you whittle down life to its barest essentials. Thoreau found he absolutely could not live without a volume of Homer in the original Greek. For me, it was really, really straight bangs.

January 3, 2009

THE LAST NIGHT, 10:17 p.m. As I write these words, a mere two hours from the start of Operation Hellhole, a full-scale media binge is under way. The girls are stalking a newfound hottie through the many mansions of MySpace, goddaughter Maddi (here on holiday) is taking nourishment from her Sidekick in gulps, and Bill sits becalmed and, I fear, benumbed before his computer game. It's that hi-def gladiator one again: enormous men with tiny heads, slavegirls in push-up bras. There's a whole lot of smiting going on.

A bit of grumbling on the way home from Gracetown today, but it's hard to fight properly after two weeks of holiday. On arriving home, Anni found the strength for one final (or so I hope) tantrum. "What you're asking us to do is not fair. You're not even asking. You're just telling!"

I "acknowledged concerns" because I thought it was a good mediation strategy, but also because she's right. She announced dramatically that she would "have to move to Dad's—but I resent it!" All sparked by my refusal to delay The Experiment till tomorrow afternoon, thus allowing her to dry and straighten hair after swimming in the morning.

Am not without sympathy, but implied strongly that the decision was out of my hands. Have given impression that the mains power is "being shut off" (i.e., by the electricity company) rather than admitting that I am turning it off myself. I am a coward. But I am an experienced coward. I know the kind of pester-power this could generate. Like, enough to light up the whole frigging neighborhood.

Made peace by reminding her she was free to straighten at friends' and neighbors'.

"But, Mum, wouldn't that be against your own rules?"

Not at all. At least, I don't think so. Hang on. Don't these people realize I'm just making it up as I go along? You know, like parenting itself.

Had planned to discuss everybody's expectations-slash-forebodings of the Big Boring Journey ahead, but chickened out. Am afraid asking them to think ahead in too much detail could be disastrous. They have been remarkably, uncharacteristically acquiescent so far. We want to keep that thought.

For myself, I anticipate missing the most:

- My straightener (LOL)
- My iPhone
- Microsoft Word (Because frankly, my writing hand has already had it)
- Google!!
- *The New York Times* online
- Pretending I don't really live in Perth

I went to the discount hardware store for supplies on the morning of January 4, trying to remember Thoreau's words: "Simplify, simplify!" Okay, it wasn't that much to remember. But it was harder than it sounds because I actually love hardware, and the more complicated the better. I can spend hours cruising for DIY stuff: shelving, knobs and fittings, power tools. Picture-hanging tackle is a particular weakness, and so, for some reason, is the fixative aisle—you know, glues, pastes, clamps, fillers. (Please don't tell me that's a metaphor.) As I wended my way toward LIGHTING, past ramparts of storage bins and leaf blowers, and down a seemingly endless aisle of hose fittings, I recalled Thoreau's advice: "Let your affairs be as two or three, and not a hundred or a thousand. Instead of a million, count half a dozen; and keep your accounts on your thumbnail."

Clearly, the man had never gone shopping for flashlights at Bunnings. Lantern-style, high-beam, fluorescent tubed, pocket-sized, solar-powered, waterproof, touch-sensitive, even head-lamp style (just the thing for reading *Walden* in a mineshaft). I spent a good forty-five minutes fretting over my choices, but in the end I think Thoreau would have been proud. Or at least not entirely disgusted. I saved my receipt—because Thoreau saved his receipts, and printed them, which is how we know exactly how much he spent to build his humble one-room house, down to the last nail and board ($28.12½). For what I spent on six lanterns, a box of candles, some large matches, a lot of batteries, and an ice chest ($240.81), Thoreau could have built a convention center.

On the way home I stopped at the deli for the largest bag of ice I could find ($3.75), which I used to fill the ice chest. I stocked it with the essentials: milk, cheese, eggs, a bar of Lindt 70 percent cocoa solids dark, a cucumber, and a pretty decent Margaret River Semillon Sauvignon Blanc. ("I believe that water is the only drink for a wise man," Thoreau had written. Yes, well. I am a woman.)

It was midmorning by now. I sat at the kitchen table and, watching the curtains stir in a weak sea breeze, reflected happily on all the chores I could not possibly get done. On a normal morning, if I wasn't at work toggling between my e-mails, my sound editor, the voice mail on two phones, and my customary six open tabs on Internet Explorer, I was at home toggling between the vacuum, the iPhone, the hair straightener, and three Word documents in varying degrees of undress. I've always considered mornings to be my most productive time. Thoreau did too, but in a somewhat different spirit. While he was at Walden Pond, he wrote, "Every morning was a cheerful invitation to make my life of equal simplicity, and I may say innocence, with Nature herself." Glancing at the $75 pile of batteries on the kitchen table, it occurred to me that simplicity was deceptively complex. The stillness was good—now that I'd experienced perhaps twenty minutes of it—but it was also, frankly, just a tiny bit spooky.

It was Sartre, I believe, who observed so gloomily that "life is elsewhere" (and he was living in *Paris* at the time). He was wrong, of course. Life is never elsewhere. And convincing yourself otherwise— that you are fate's victim, or prisoner, or terry-cloth hand puppet, even—only underscores the point. That's one thing my not-entirely-self-imposed exile Down Under has taught me.

When I told my Australian boyfriend I'd follow him to the ends of the earth, I had no idea he'd take me so literally. We were in graduate school at NYU, finishing our respective doctorates, when Ron, a Sydneysider, was offered a job in Perth, Western Australia. If it hadn't been for that job offer, we probably never would have gotten married. We certainly would never have settled in Perth—a move that, from my perspective as a New Yorker (even an idealistic New Yorker, helplessly awash in the dopamine-scented haze of young love) felt more like a transfer to the lunar surface.

When we divorced three years later, it was hardly what you'd call a bolt from the blue. He was an Anglican priest who enjoyed golf, tennis, and vintage port. I was a kick-ass feminist intellectual who excelled at chain-smoking. It was never gonna happen. Our geographical differences were equally irreconcilable. He was overjoyed to be back in his own country after four years in New York. As for me, as much as I appreciated the clean air and gorgeous beaches, I knew I could never in a billion years call Australia home.

That was twenty-four years ago. LOL.

In the meantime, I remarried—a doctor this time—and had three babies (but not necessarily in that order) with alarming alacrity. The kids were four, two, and six months old when we broke up. My first divorce had been sad but amicable. This one was a conflagration. With an Australian as the father of my children, my chances of moving back to the United States were now as remote as Perth itself. I'd followed my heart to the ends of the earth, all right. And now I was stranded here.

That was fourteen years and several lifetimes ago. I look back at those days when I wished my ex-husband would get run over by a bus and feel pretty ashamed. I realize now that migration to a third-world country would have been fine. Kidding, people! He was, and very much still is, the father of my children. I remind myself of that every day of my life. The kids love him, and he loves them, and whatever the differences between us, that should be good enough for me. It's not. But at least I recognize that it should be. Like that half-done quilt I've had stuffed into a bottom drawer since my eldest started middle school, I'm working on it.

Death, it is said, concentrates the mind wonderfully. So, too, does having sole charge of three kids under five. In the early years, like many another single parent, I clung to my professional identity like a baby to an umbilicus. I started writing a weekly newspaper column about my kids. I started writing books about gender and family life. And I started planning. (Up until this time, I decided, I'd lived in the here and now perhaps a little too successfully.) I formulated a five-year plan to get us stateside. Six years later, I revised that to a ten-year plan, and then a fifteen-year plan. In the meantime, the lioness's share of my energy went to parenting my children: one sticky, sleep-deprived, extraordinary day at a time. I reminded myself that if home was where the heart was, then by any real reckoning I was already back in Kansas. Some of the time, I even believed it.

All this backstory matters, because, in a funny kind of way, being . . . well, *cut off* has been such a central theme for me. You might almost say that going off the grid has been the story of my adult life. It has certainly served as the wellspring of my IT attachment issues. Simply put, digital media have made it possible for me to live in two places at once—Australia and America—in a mind/body split so sustained and ambitious, it makes Descartes look like a cheese grater.

When I arrived in Western Australia in 1986, an airmail letter from New York took two weeks to arrive. ("You needed special onion-skin paper, special stamps, special stickers," I tell the kids in a quavery voice. "Seriously, we're talking one step beyond sealing wax and a signet ring.") Trans-hemispheric phone calls were like Woody Allen's joke about the meals at the Catskills resort: They were terrible and you never got enough of them. Annoying three-second transmission delays ensured plenty of awkward silences and inaudible cross-talk, and an eerie, swishing echo that lent conversation all the intimacy of a Jacques Cousteau special. In fact, "conversing" was a misnomer. Basically, you gurgled. And if the party you were speaking to recognized it as *your* gurgle, you were satisfied.

There were even bigger problems than the four-week turnaround to find out how your two-month-old nephew was (now that he was your three-month-old nephew). The *New York Times* crossword puzzle, for example. Like Rapunzel pining for rampion, my craving for it grew so intense that my worried husband was forced to forage for it—in one case, in the U.S. Consulate General's office on St. George's Terrace. It's a kindness I have never forgotten.

Today, *The New York Times* is my homepage. I read it on my iPhone on the train to work. I do the daily puzzle then, too, if I feel like it. But there's really no rush. Because as a premium crossword member, I have access to more than four thousand other puzzles, and solutions, from the *Times*'s archives.

How can I begin to explain how such innovations have changed my life? I stream NPR's *Morning Edition* live in my bedroom (albeit in the late evening the previous day, owing to deep time-zone weirdness). I listen to a gazillion U.S. podcasts through my car radio—including almost every program broadcast by my "home" public radio station, WNYC. WNYC! On the Kwinana Freeway, heading south over the Swan River. Past the suburbs of Dog Swamp and Innaloo, and the aspirationally titled Perth Entertainment Center!

I can talk to my family via e-mail, Facebook, IM, or Skype, instantaneously and in real time, whenever I want to. With webcams, we can pretend we're all in the same room, let alone the same hemisphere. (Which, come to think of it, is maybe why we don't do it that often . . .) I can call my sister's cell phone on Long Island from my cell phone in Fremantle. I can order American music, books, and DVDs direct from Amazon. I can download American television direct from iTunes—occasionally going to heroic technological lengths to do so (by purchasing a U.S. iTunes gift card on eBay in Australian dollars from some guy in Spain, as I did recently).

Before the Internet, getting books from the States—or even from Sydney or Melbourne—was a long-haul operation. A shipping delay of three to six months was standard. No exaggeration, the words "sea mail" make me nauseated to this day. Today, I can get an e-book as fast as anybody, anywhere—in about a minute. The first time I tried it, I literally wept for joy.

For an information junkie in exile like me, the dawning of the digital age has been like the arrival of a rescue ship.

Even on the occasion of my first encounter online—back when "websites," so-called, consisted largely of dense pages of alphanumerics—I knew instantly that my world, and by extension THE world, had shifted on its axis, irreversibly. The chill that went down my spine was the same shock of recognition I felt the morning I met my husband, the day I watched my son play his first game of water polo, the moment I beheld my firstborn's face: that there was life before this moment, and then there would be life after. And I know that sounds pretty dramatic. But . . . that's the point.

After twenty-four years, two husbands, and more changes of

hairstyle than I care to count, I am still a New Yorker who's just passing through. It's a cultural identity as stubborn as any birthmark, which no amount of exposure to the relentless Western Australian sunshine can fade. For me, information technology is so much more than a means to an end. It's a hotline, a lifeline, to my Real World. The one in which I cannot physically or directly participate. The one that exists Up Over, half a world beyond the impossible blue of the Indian Ocean and the breeze sweeping in through the kitchen window.

As I swatted a blowfly dead—blowflies are to the Western Australian summer what blackheads are to being fourteen—it occurred to me that the biggest challenge of this whole project might have to do with relinquishing that ostrichlike delusion: that burying my head in information and entertainment from "home" was just as good as actually being there.

But for now, there were many smaller decks that needed to be cleared. At the moment, we were screen-free in theory only. In reality, it was a case of media, media everywhere and not a drop to drink.

The laptops were no problem to pack away—I stuck all three of them in a filing cabinet, under H for hibernation—but bedding down Bill's beloved gaming PC took a bit more grunt.

The Beast, as we called it (and the towering gray chipboard enclosure in which it brooded), had been allowed to become the focal point of the family room. Now it sat slightly askew, covered in dust and discarded peripherals, like a ruined monument from some long-vanished race of teenage barbarians.

A massive monitor sat serenely in the center of it all surrounded by offerings of half-drunk water bottles and crumpled candy wrappers. I rolled the table out from the wall, uncovering a snake pit of cobwebby cords, cables, and connectors. Also a physics textbook. I spotted a hank of what I feared was human hair but turned out to be

a tumbleweed of Rupert's undercoat and some pencil shavings. I took
a photo, just for old time's sake. And then I rolled up my sleeves.

Over the next two hours, I unplugged and coiled up chargers for
a bewildering profusion of digital drek: two mobile phones, one Nintendo DS, a PS/3, an iPod, two vintage Game Boys, several thumb
drives, an external hard drive, and a digital camera. Three-quarters
of these devices were missing and presumed dead, but I saved the
chargers anyhow. Because they might show up one day, like Bo Peep's
sheep or some deadbeat dad—and also because I am, alas, a hoarder.
To be honest, it was all I could do to toss out the tumbling tumbleweed.

The rest of the gear I dusted, coiled, and stashed at the back of the
old TV cabinet, next to the Barney videos. (Yeah, well. We might want to
watch them again someday, okay?) The Beast itself would be leaving for
a working holiday, a sort of whistle-stop tour of Bill's friends' bedrooms. I
packed it a little bag—a USB cable, some DVDs, and a thumb drive, just
in case it got peckish—and lugged it to the front door to await pick-up.

I'd always worried that being a single-parent family somehow put
us at greater risk of information abuse. But no, according to the
Pew Internet & American Life survey. The latest figures show that
two-parent families with children have the highest technology
concentration of any household type. Today the average eight- to eighteen-year-old shares his home with two computers, and 84 percent of
American children have a home Internet connection. So, all up, our
tally of one desktop and three laptops—a networked computer for each
member of family—is maybe not typical, but it's totally un-unusual.[1]

The televisions were next on my hit list. I aimed to rub them out by
the time Bill splashed down from training. (If there's a single thing Bill
is more passionate about than The Beast, it is being a water-polo goalie.
With no digital distractions, it was a fair bet that the pool would become
his new default setting.) According to the Kaiser Foundation, the

typical home with kids has 3.8 televisions sets, 2.8 DVD or VCR players, and 1 digital video recorder. Thirty-seven percent of such households also lay proud claim to a video screen in the family car. Among
all eight- to eighteen-year-olds, 71 percent have a TV in their bedroom,
and half also have a video game player and/or cable TV. In this, as in so
many other respects, we were aberrant, only this time it was in a good
way. We had only two TVs, not counting the battered black-and-white
twelve-inch Sussy'd bought for five bucks at a craft fair when she was
nine. ("Interesting handiwork," I'd sniffed at the time.)

Figures from the Consumer Electronics Association show that
87 percent of American homes have cable or satellite TV. In Australia, less than a quarter of households do, and we had left their ranks
about six months before The Experiment began. Personally, and I feel
a little weird admitting this in public, I don't really *do* television. It's
not a moral issue. I'm just like one of those natural vegetarian types,
who freakishly happens to prefer flaxseed to steaks.

I wasn't always like this. When I was a kid, my whole day was structured around my favorite shows, and had been from the moment I first
clapped eyes on *Romper Room*, circa 1961 (think taffeta party dresses,
miniature sports coats, and patent leather pumps for *all my friends
today!*). But something happened to my television viewing around the
time I went to college. Like my virginity perhaps, it just sort of with-
ered away. When I had kids, it seemed to disappear altogether.

As a single parent, I found that the last thing I craved at the end of the
day was noise. Of any kind. I couldn't even listen to *music* for years. In my
frame of reference, television was strictly a child-deflection/distraction/
diversion device. They'd watch *Play School* or *Thomas the Tank Engine* or
Lady Lovely Locks so that I could get on with cooking dinner and match-
ing socks, or meeting deadlines and—very occasionally—a man.

In fourteen years as a single parent, I have dated only three men,
yet ended up living with two of them. Separately, I hasten to add! The
second of those partners, a one-time Australian Olympian, arrived

with a twenty-four-hour sports channel in his wake, as part of a handsome subscription television package, and the impact on the family's television viewing habits was dramatic.

Sussy got in touch with her latent American roots and discovered sitcoms, and the dopier they were, the better she liked 'em. (*The Suite Life of Zack & Cody*, for example, featuring eleven-year-old identical twins who make their home in a luxury hotel, where their single mother performs a nightclub act. Enough said?) Bill dabbled briefly in World Wide Wrestling, before veering off in the direction of *Family Guy* and *The Daily Show* and a Dantesque loop of Australian Rules Football. Anni, interestingly, watched nothing at all—ever. She felt about our new satellite capability more or less the way she felt about the new partner: "Not a fan," in her own terse words.

The relationship broke up, but the Lifestyle Channel lingered on. It wasn't till I'd reached the acceptance stage of relationship bereavement that I finally had it disconnected. By that point, the children had discovered the illicit joys of downloading. Holed up in their bedrooms with the latest episodes of *The Hills* or *The Secret Life of the American Teenager* or *Naruto*, the family-room television seemed sooo Edward R. Murrow. It was time to move on. We hauled the big, old television to Bill's bedroom, and I bought a cheap "fat screen" replacement set, much to the children's humiliation. It was under the load of this embarrassment that I was now staggering toward the garden shed.

I set it down carefully atop a Styrofoam bodyboard, as if it were going on a long sea journey. Bill's TV, which easily weighed as much as I, or for that matter the shed, did, was going to be more of a challenge. But I was desperate to get rid of it before he got back from training. It may sound ridiculous, but he was so attached to it—and proud of it, almost, the way boys *are* proud of their machines and gadgets—I felt guilty about taking it away from him. Mean, too. I'd been so dismissive of the friends who'd said, "Are you sure you want to do that to your children?" Now, for the first time, I wasn't so sure.

What would I do if Bill suddenly snapped? If he refused to part with the TV and reneged on his promise? I'd be heartbroken. Not just for the plans I'd made (elaborate though they were). But for his sake. For all their sakes.

I wanted my kids to experience this—and I wanted it in exactly the same way and for exactly the same reasons that I wanted them to travel overseas, or practice yoga, or learn a foreign language, or take sailing lessons: to enlarge themselves. To discover themselves. To become human beings more fully alive, in the *Walden*esque words of Saint Irenaeus. My children have lived in Australia all their lives. But in important respects they have been raised elsewhere, in the supranational city-state prophesied by Marshall McLuhan in the 1960s, and made flesh by Bill Gates and Steve Jobs in the 1990s. Digitopia. Cyburbia. The Global Village. Call it what you will, it is where they live now. My children happen to be dual Australian–American citizens. But first and foremost, they are Digital Natives—just like yours.

When I encountered the term "Digital Native" in John Palfrey and Urs Gasser's 2008 book *Born Digital*, I didn't even need to read the definition. I knew instinctively what it meant and to whom it referred. The term was coined by game designer and self-described "visionary" Marc Prensky in 2001 and refers to "the first generation born and raised completely wired," in the words of Palfrey and Gasser.[2] That means Anni, Bill, and Sussy, all born between 1990 and 1994, definitely qualify for membership. I don't.

Despite my hard-won technological know-how, and even though I can do some online tasks better than they can (well, one anyway: If there was a TV show called *So You Think You Can Google?*, I'd be an overnight sensation), I am by definition a Digital Immigrant—and if you were born before 1980, so are you.

No matter how tech-savvy we Digital Immigrants become, we betray our Old World origins at every turn. Starting with reading the instruction manual, which is a dead giveaway. When a new technology arrives in the home, the Natives don't need to set out on a humbling search for the "On" button. They just know, as though they've been fitted with an auto-detect device their elders and inferiors have only read about in the IT pages of their sad little newspapers. A new application is the same. Immigrants such as you and me wade conscientiously through the documentation. We do the tutorial. We register for online support. In short, we approach every new media experience—from Twitter to TiVo—as if it were a digital disaster waiting to happen. We respond by lining up a walking stick and a wheelchair, just in case.

The Natives find this hilarious. They are no more frightened by new media than they are by a new pair of running shoes. They just jump right in and start sprinting. While we're struggling with setting the time and date, they've shot a music video, customized the ringtone with seasonally appropriate sound effects, and changed the wallpaper to a close-up of the pug relieving himself.

We Digital Immigrants work hard on our second-language skills. But we still speak "Download" haltingly, and with thick foreign accents. We say "write" where we should say "post"; "page" where we should say "profile." We balk at using "friend" as a verb. We "dial" our phones and "look up" facts. We forget the word "avatar," if we ever knew it to begin with. And we appeal constantly to our kids to translate for us. ("Hey, guys! Cousin Linda just threw a barrel of monkeys at me on Facebook. Please tell me this is a good thing.")

Like most others of their tribe, my own Digital Natives regularly traveled to foreign lands, albeit in short hops, even before the Winter of Our Disconnect descended. They did read books (every once in a while). They played music and sports, and went shopping. They indulged in the odd bout of extreme hair coloring. Lord knows, they would always press "pause" to eat—grazing pretty much constantly, like a small but attractive herd of goats. And, of course, no matter what other balls they were keeping in the air, they always seemed to find time to annoy one another.

The whole point of The Experiment was to send the Digital Natives on an *extended* trip abroad. A sort of family holiday back to the Old Country. An immersion experience, if you will, in the culture of their forebears. In time, I dared to hope, they might even adapt some of their quaint folkways for use back home in their native land.

But such lofty dreams would have to wait. For now, I had a more modest goal: to move the goddamn television already. I repaired to Bill's bedroom, located at the very back of the house, for a site visit. It had been a long time since I'd surveyed the Crack Den in any objective way. Looking around me now, I could understand why. The TV was the size of a refrigerator box. The room was the size of a refrigerator. To say it dominated the space was an absurd understatement. The good news was that the television sat on a dresser that was just at waist height. Ergonomically this was propitious because it meant I didn't need to bend my knees to lift it. I could easily wrap my arms around it, if only I were a gorilla. As it was, the thing was as bulky and unbudgeable as a husband on a couch. There had to be another way. Maybe a wheelbarrow? If so, I'd have to work fast because Bill was due back any minute.

I made haste to the shed, half expecting to find Lucy and Ethel crouching behind the fertilizer. (Over the next six months, I would experience many such moments. If I'd ever secretly wished my life were being filmed live in front of a studio audience—and trust me, I hadn't—it would have been deeply gratifying.) Lined with an old horse blanket, the wheelbarrow made a serviceable rickshaw. I wheeled it around to the back entrance, just a step from Bill's bedroom door. I even managed to grip the TV and heave it off the dresser. What I couldn't do was turn or walk. Or breathe. I was rooted to the spot, frantically wondering what to do now, before my forearms unhinged like Lego pieces, when Bill came charging down the hallway, a pool towel flapping, capelike, around his ankles.

"WTF, Mum!"

Oh God. I hate it when he uses bad abbreviations.

"What the hell do you think you're doing?" he repeated crossly. "Let *me* do that."

At that moment, I was so moved by his unexpected affirmation, I practically found the strength of ten menopausal mothers. To me, it meant Bill was in it for the long haul. Metaphorically and otherwise, that took a huge weight off my back.

In the end it took all we had—and he's a six-foot-one fifteen-year-old with shoulders as broad as an ax handle—to wrangle the TV out of the back door, into the klieglike glare of the midday sun, across the grass, and under the orange tree to the suffocating blackness of the shed. Somehow or other, Bill balanced it on top of the other set, crushing the boogie board and an index finger (mine). He covered it lovingly with a blanket and trudged manfully back inside.

From my perspective, the next two weeks passed—or wafted, really— as if in a dream. "That's because it was so dark you couldn't *see* anything, Mum," Anni reminded me tartly. But even so, Blackout Bootcamp proved to be one of the most serene and transforming periods of my entire adult life. A cynic might say it had something to do with the fact that the lights were off but nobody was home. There was a kernel of truth there, no doubt about it.

Sussy did indeed move out entirely—taking her suitcase and her MacBook with her. "I really think it's time I spent more time with Dad," she explained again earnestly, and not entirely convincingly. I wasn't happy about it. But she probably did need to spend more time with her father, whom she'd tended to see only sporadically in the last few years. He lived in a country town about an hour's drive south, but had a pied-à-terre just down the road, where he stayed during the week. And if Sussy really believed that home was where the MySpace was, here was a perfect opportunity to test the hypothesis.

I tried not to feel "blocked"—to use the language of social media—but I wasn't always successful. There were many times over the years when I'd felt reduced to a kind of glorified service provider to my children, but this was taking it to a whole new level. I told myself it would be a learning experience for all of us, and kept combing through *Walden* for consolation. "A man is rich in proportion to the number of things which he can afford to let alone," I read. Yes, but not your *daughter*, I couldn't help thinking.

Although they were more subtle about it, Anni and Bill also took a fugitive approach during these early weeks of tepid milk, hot sheets, and ice-cold showers. (It had been a shock to discover that our gas-powered hot water system required electricity to ignite. When I figured this out on the second day—there had been enough reserve hot water to see us through the first—even I started to wobble. If the weather hadn't been so sultry, The Experiment might have dried up there and then.) There were *a lot* of sleepovers at friends' houses. Less predictably, there were a fair few incoming sleepovers as well.

The kids had plenty of friends who'd been back and forth like blowflies to Europe and North America. But nobody, *nobody* had been on a power trip like this one—not intentionally and in their home. The Harry Potteresque lanterns weren't the only draw card. So too was the opportunity—if you can believe it—to play board games. I hadn't anticipated that Blackout Bootcamp would have such strong novelty value among the been-there, done-that crowd.

The first night without power, Sussy, Maddi, and I kicked it off with a round of ImaginiFF. And *they* initiated it. ("Imagine if we could turn on the fan," Sussy quipped.) I tried to remember the last time any of my children had asked me to play a game. Not counting mind games, it had been years. Sure, we'd played poker and Yahtzee when we were on holiday with other families—especially at Rottnest Island where, like Gracetown, the lack of technological amenities was world class. And of course I'd played lots of games with them when they were little. (Like a good feminist, I'd taught them that the player who gets stuck with the Old Maid is the winner.)

But everybody had grown up to be so damned competitive, I'd purposely steered away from anything that involved winning and losing. They still arm-wrestled most mornings over who was going to ride in the front seat of the car. I certainly wasn't about to take my life into my hands and play Monopoly with these people. The Experiment would mean we had less choice about whether or not to cooperate—like that Hitchcock film where the people are stuck in the lifeboat and they all have to pull together or die of exposure.

A few nights in, Bill's friend Pat slept over. It seems he'd had a huge fight with his parents and his brother about computer time. LOL! "Did Bill tell you we have no power here?" I asked cautiously. "Fine with me," he'd growled. "I'm over it."

The last time Pat and Bill had had a sleepover, Pat had brought his desktop computer along with him—strapped to his bike like a large child. This was not unusual. The friendship was really a foursome: two boys and two PCs. No matter how often Bill explained it to me, I could never quite figure out why this was necessary—like, if they were playing games, couldn't they just take turns?—but it was something about battling each other in real time. Frankly, it sounded a lot like being married.

This time, Pat brought a toothbrush and a book instead. "Wow. Pat can read?" Sussy hissed. (She was home on weekend furlough.) That night, I went to say goodnight and found the boys sitting up on Bill's bed, side by side with their Coleman lanterns and their books: *Harry Potter and the Prisoner of Azkaban* and—get this—*The God Delusion*.

So Dawkins was wrong after all, I reflected as I tiptoed down the hall. There really *is* a God.

―――――

January 5, 2009

Peaceful, almost Zen-like atmosphere in house today.

Duh. No one home.

Seriously—the quality of the silence has changed. It's thicker, more meditative. The buzz is gone. It's good.

Cleaned fridge and found two gel eye-masks buried under a drift of pecorino. Gross! Made chicken curry for dinner. Girls (Anni, Maddi, Suss) returned around eight p.m. and screamed with laughter to find Bill and self eating it on verandah by kerosene lamplight. "Creepy!" they cried. (A: "Mum, this is the dorkiest idea you ever had.") But dutifully took up their lanterns and went inside.

Snuck down hallway later to find all arrayed in A.'s bed surrounded by lanterns and mags. Suss reading novel titled *No Fat Chicks*. Others discussing "fear of intimacy."

Bedrooms terrifyingly slatternly at present but thankfully cannot see much detail.

January 6

Wrote column in longhand, just like this diary. Painful to hand and head, big-time. Harbor no ambivalence whatsoever re: MS Word. It rocks. Sigh.

Bought bunch of new pens, notebooks as treat to self. (Sharpie ultra fine points, permanent, and spiral bound nbks with sober but elegant black covers.) Spent foolish amount of time trawling well-lit, climate-controlled aisles of OfficeWorks. Can see myself starting to abuse stationery if not careful.

Keywords wish list (i.e., stuff I wanted to Google today): 1—natural diuretics, 2—"French justice minister" AND pregnant, 3—Perth New York airfare cheapest, 4—cause of death, HD Thoreau.

Read every blessed word of newspaper.

Bill rode bike to Vinnie's and just called to ask if he could stay the night. Evidently The Beast still roams, seeking whom he may devour.

January 7

Have totally settled into *Walden*-worthy routine now. Spent morning at South Beach, snorkeling, snoozing, and rereading Thoreau. Home for grilled cheese cooked in frying pan. (NB: Have discovered how to make toast over an open flame. Spear bread with long fork, wave in circles over gas ring. Avoid observing self in rangehood.)

Definitely eating strangely, out of all routine. Today: ½ almond croissant, 2 mangoes, 1 cheese sandwich, 1 glass wine, 1 grapefruit soda, 1 Kit Kat. Thoreau would gag. He did have some pretty odd cravings himself, though. "I caught a glimpse of a woodchuck stealing across my path," he wrote, "and felt a strange thrill of savage delight, and was strongly tempted to seize and devour him raw; not that I was hungry then, except for that wildness which he represented." Interesting. Felt much the same way about the Kit Kat.

January 8

Near 40° today—same for tomorrow. Cannot Google "metric converter" for precise Fahrenheit equiv but know it's over a hundred. (Weird how after twenty-three years of metric I still feel this need. "Ten kilometers? Yes, but what's it really?" I always want to know.)

Taking many walks, despite heat. V. strange without iPod. Normally have certain podcasts for certain routes. (*Wait, Wait Don't Tell*

Me on Lefroy Rd heading west. *This American Life* on Dog Beach, etc.) Unplugged, am trying instead to focus on surroundings. Taking new routes so I have something interesting/new to observe. Figs, e.g. They are everywhere in this neighborhood! Am trying one from every tree I pass. Thoreauvian? Maybe. Fattening? Definitely.

Hair issues continue to challenge. Have experimented with low-tech straightening—i.e., combing wet bangs straight back and pinning them as if skull were a giant roller. Sad really.

Do look, as feared, like bag lady. But at least have acquired good tan. So perhaps a glowing bag lady . . .

January 9

Tension.

Fought with A. about dishes, credit-card charge for overdue books, and general princess attitude. Returned home from clubs last night at TWO A.M. Silently handed her and friend a lantern and stalked back to bed. ("Sorry about my FREAKY HOUSE, Laura!" she called out.) Always blaming The Experiment for everything undone, disorganized, and dysfunctional in her life right now. Literally ask myself on a daily if not hourly basis where I have gone wrong. Powerlessness? Don't get me started.

January 10

Bad night: hay-feverish and disturbing dreams. I want my, I want my, I want my NPR!

Explained to B.'s coach about The Experiment and was startled/slightly freaked out to see his eyes misting over. He gave Bill a lecture about how fortunate he was to be having this experience . . . how he would remember it his whole life AND BE GRATEFUL TO HIS MOTHER (my favorite part). Said he would love to do the same at his house. Bill dumbstruck. Looked at me with something bordering on respect.

Anni apologized for yesterday and we did a big, cathartic clean of her room. ("Simplify, simplify, honey. Like, why have a dozen empty cans of Diet Coke when two or three will do?") Said she'd told friends about The Experiment and was surprised how many thought it was cool.

Did laundry by hand, btw. Surprisingly pleasant (just a couple of summer tops and underwear—can't imagine doing linens or, God forbid, B.'s football stuff). Remembered my grandmother's wash-board and how she washed her "smalls" in the kitchen sink every morning and hung them out to dry first thing. Felt v. virtuous and carbon neutral, pegging it all out like some fifties housewife. Shame about the hair though.

January 11

NEW KITTEN!!! Decided we needed a handheld toy after all. Hazel! So adorable. A. & B. picked her out of lineup from Cat Haven. She is afraid of Rupert, even though he is about as fierce as a Persian rug.

Eleven p.m. Girls talking and reading by candlelight, in clean, aired room, not glued to Facebook in zombie-ish oblivion to surround-ing chaos. They are tired—as they should be at this hour—not wired.

January 13

Have decided dishwasher is hugely overrated. Not really a time-saving device—more a time-delay device. Its function surely ¾ aesthetic—i.e., removing dishes from view. A dark kitchen does the same job instantaneously.

January 15

How to manage time (drop-offs, pick-ups) without cell phone? First attempt today, as needed Bill's help to pick up bed. (Offered him queen-size bed to compensate for loss of TV. Will have literally no room to swing a pug in there now, but whatevs . . .)

Complicated logistics! Me at ABC in Perth, B. in Fremantle. Devised plan for B. to take bus and then train to Subiaco, where would meet him at station at 3:30 p.m.

V. anxious, as plans like this usually need a dozen texts back and forth to confirm, rejig, reconfirm, and re-rejig. ("Missed bus," "Can't find SmartRider," "OK to meet fifteen mins later?" etc. etc.)

And guess what?

Bang on time. When I saw his head appear in the crowd on the up escalator, practically punched fist in air (but knew that would make him go down again!).

It's like the old days, when people just *showed up*.

January 16

Am so over that damned ice chest. Wine bottle tipped, and bouquet of cut-rate Sem Sauv infuses all provisions. Nasty scratch on countertop too.

Went to Bill's room this a.m. and caught him playing with SHADOWS. Besotted with Hazel, as are all. (Suss put out she was not in on selection. Sat on rug for hours playing hide and squeak.)

Missed iPhone painfully when had to wait twenty mins with B. at Thai takeout joint. In desperation, played Mr. Squiggle and Scattergories instead. Pretty fun. A notepad makes a decent handheld game actually. Had forgotten that.

January 17

10:30ish—last night of powerlessness!

Anni literally counting the minutes till midnight. She is reading *Dating Up* by lantern light, surrounded by candles, Hazel dozing on chest.

January 18

I let there be light. Also power.

And it was very, very good.

B., capering around kitchen: "Electricity is awesome! Electricity is awesome!" Switched on microwave just to hear the beeps and danced away again.

A: "I don't care if I NEVER have my computer, as long as I can read in my bedroom with the fan and the lights on."

S. "visiting" again today. "I miss Hazel," she explains. Uh huh.

The Beast returned from its wanderings. Lies defeated in the hallway, as if to say, "Look on my graphics capability, all ye mighty, and despair!"

Straightened hair slowly and with deep, soul-satisfying enjoyment. Baked muffins (corn).

Electricity was still awesome the next day, and remained awesome for many weeks to come. Yet as withdrawal periods go, we got off pretty lightly, I think. No one broke out in a cold sweat or hallucinated visits from the ghosts of playlists past. But the darkness drove us apart in some ways. There was Sussy's departure, most spectacularly. And frankly we all sought our own places of refuge: in friends' houses or movie theaters, at the town pool, or in Wi-Fi-enabled cafés on Fremantle's cappuccino strip. My own little haven was a certain rocky outcrop on the northernmost corner of South Beach, where I spent hours reading *Walden* and hiding my frizzy bangs. And that was okay. After all, that was a major part of the plan: to pry us out of our respective digital cells and into that e-mail-free zone called "life itself."

Anyhow, as Thoreau himself reminded us, "How could youths better learn to live than by at once trying the experiment of living?" Now that the lights were back on, the real fun (if that's what you want to call it) was about to begin.

Boredom for Beginners

Don't just do something. Stand there!
　　—ANONYMOUS

That very first morning, I awoke to the sound of birdsong. "Shiznit!" I muttered. (Living with three teenagers isn't a vocabulary builder.)

Leaning across to turn down the volume on the CD player—that frigging Balinese relaxation CD was getting on my nerves—I suddenly realized. OMG. The birds were live. It was everything else that was dead: my alarm clock, the fan over my bed, the lights, the appliances, and every medium and device that we owned. From the family room, the usual sounds of a Sunday morning were conspicuously—almost creepily—absent.

No MTV. No *Video Hits*. No burst of antiaircraft fire or charge from a sniper's bullet. No pinging or bleeping or—barring the magpie—chirping of any kind.

To be honest, part of me was weirded out. But another part of me, possibly my middle ear, trembled with something that might have been joy. I recalled Thoreau's words about the healing power of quietude: "There are many fine things we cannot say if we have to shout." At that moment, I felt such a rush of gratitude and certainty. I knew, deep down in my soul, that what we were doing was right, and long overdue. I lay there, soaking up the silence, practically pillowed in it, when there was a knock on my door.

"Mum!" the voice thundered.

I sank under the covers, cowering like some large maternal ostrich.

"Muuum, I'm bored." It was Bill, a towel slung over his shoulders, demanding to be taken to the water-polo pool. It seems he'd spent "like an hour" already rereading Harry Potter, and his screen-free options were now exhausted. I'm sure there were many fine things I could have said without shouting. I just couldn't remember what they were.

One hour down, only 4,379 to go. Oh, boy.

After we'd packed away our technology, I spent hours disentangling stuff, getting uncomfortably up close and personal with the snarl of cables, connectors, and tufts of pug fur that lurked behind our work-stations. Bleugh! Who knew it was such a mess back there? When the job was finally finished, I felt—to use my daughters' favorite verb—cleansed. Over the next six months, I spent many more hours disentangling my ideas about our technology, separating out assumptions from observations, unraveling articles of faith (and hope) from inconvenient truths. Bleugh! Who knew it was such a mess in there?

Boredom was one of the first knots I encountered. And it was boredom that got me thinking about Sir Don Bradman.

As every American schoolchild knows the story of George Washington, every Australian kid learns the story of Don Bradman, a little boy from the bush who didn't grow up to be prime minister, but something much, much more important. Captain of the Australian cricket team. To be fair, Bradman wasn't just any cricketer. He was the God of cricket. The greatest and most beloved batsman in the history of the game.

Bradman was born in 1908 and raised in country New South

Wales—the original "back o' beyond." Legend has it that the young Don had no cricket ball or even a bat, but only a golf ball and a stick, which he batted devotedly and not a little compulsively against a rainwater tank in the long, hot, dusty afternoons.

"At the time it had no meaning," Bradman later admitted. "I was just enjoying myself. I had no idea that I was training my eyesight and movements." By the age of twelve, Bradman was declared a child prodigy of cricket. His career batting average of 99.94 is widely acknowledged as the greatest statistical achievement in any sport.

All because he was so lavishly talented? Or all because he was so lavishly . . . bored?

Boredom is something the Winter of Our Disconnect gave us plenty of time to ponder. It was there from the first moment I'd imagined The Experiment, one day toward the end of 2008 when I'd been having another in a long series of conversations with the back of my son's head. He was playing *Jason and the Argonauts* without a pause as he pretended to listen to me. Don't ask me how you can tell this from the back of a person's head, but trust me, you can.

It's like talking on the phone to someone who's reading their e-mail. There's a hooded quality to their voice that's kind of the vocal equivalent of a blank stare or a busy signal. I don't remember exactly what I was trying to talk to Bill about that day. It could have been anything. Whether he'd fed the dog. Whether he believed in a personal savior. Whether he planned to turn around and make eye contact before the end of the financial year. Whatever it was, the grunts I was getting by way of reply were setting my teeth on edge, like a borrowed mouthguard.

Anni was behind me, stationed with her laptop at the so-called craft table. A few years back, when its indestructible, recycled pine-plank surface was still hidden under a happy clutter of paints and scissors, glitter and smoking glue guns, the name made sense. By 2008, it had been years since the craft table had been used for anything more craftsmanlike than inserting a thumb drive, and only a few stray deposits of fossilized Play-Doh remained to mark the passing of an era.

Sussy sprawled on the couch with her Nintendo DS, I remember, playing some weirdly addictive Japanese cooking game that was all the rage with her Year 9 classmates. ("It develops skills, Mum," she'd huffed the last time I'd ventured to criticize it. "*You* try frying eggs on a screen this size!") Beside her, Rupert looked on benignly, yet with a touch of anxiety—something you can generally count on a pug to do.

"What would life be like," I heard myself mutter, "*without* all this crap?" No one looked up.

"What would life be like," I repeated, more loudly this time, "if all our screens suddenly went blank—if we just pulled the plug on the whole shootin' match?"

Maybe it was the word "shootin'" that did it. Who knows? But Bill responded with a full sentence. In fact, he almost turned around.

"It would be boring, that's what," he replied.

"IT'S . . . BORING . . . *NOW!*" I wanted to shriek.

I gazed from one set of flying fingers and glazed eyeballs to the next. They say you shouldn't shout at a sleepwalker. I decided to try a different tack.

"Maybe being bored isn't such a bad thing," I ventured.

"And maybe it is," he replied, right on cue. I asked for that.

"Anyway, what does 'boring' really mean, Bill? Like, what did people do with themselves before computers, or for that matter before Gutenberg?" There was a pause.

"As if *Police Academy* has anything to do with it," Suss muttered. Sad to say, it wasn't a quip. I soldiered on.

"Like, was everybody just 'bored' all the time, or what?"

"They probably were, Mum, but they just didn't know it," Bill responded, a bit uncertainly. He went back to hitting somebody over the head with a mace, but I could tell he was thinking about it.

Quite obviously, boredom is all *about* perception. It's a self-diagnosis, pure and simple. If you don't realize you're bored, you're not. For a few minutes all you could hear was the sound of Sussy's virtual eggs frying in their virtual frying pan.

"Because if they weren't bored," Bill added at last, "why would they have invented the computer, or whatever, in the first place?"

Ah! So it *was* possible to think and smite the enemy at the same time. Impressive.

Bill's point—basically, that boredom might be construed as the impetus for achievement rather than as an obstacle to it—made intuitive sense (and, given that he'd lent unexpected support to my own case, I was gracious enough to acknowledge that at the time). Months later, when I started to investigate the topic more systematically, I discovered how spot-on his hunch actually was. The role of boredom in encouraging innovation and creativity *is* a critical one. It's not only history that teaches us this—like the story of Sir Don and his bionic batting eye—but the evidence of our own life experience. So why have so many of us forgotten this simple truth: that motivation begins with *discomfort*—with needs that are *un*fulfilled?

"Some desire is necessary to keep life in motion," observed eighteenth-century man of letters Samuel Johnson. "And he whose wants are supplied must admit those of fancy." Which is arguably where the virtual fried eggs come in.

It's like the joke about the nice Jewish boy who still lives at home at the age of forty, and has never spoken a word to anyone—until one night at the dinner table, he suddenly says, "Could you pass the salt, please?" His amazed parents cry out in wonder. "My son! You can speak! But tell us, why have you never spoken before?"

The son shrugs. "Up to now, everything's been fine."

Twentieth-century philosopher William Barrett was getting at pretty much the same thing when he observed our primal need to see the universe as being "rich in unsolved problems." Without that perception, Barrett argued, we feel at loose ends, literally purposeless and maybe even panicky—like my mother obsessing over the laundry while she's supposed to be living it up on a Caribbean cruise ("I can't meet you at the pool till I've washed and hung out my underwear," she'd tell my dad in all seriousness) or, for that matter, my "retired" father keeping meticulous records of his personal-best lawn-mowing times. Post-Experiment, I can see that the games that once absorbed so much of my kids' hard-drive space were functioning in the same way: creating imaginary problems to solve in an existence unnaturally and possibly dangerously deficient in real ones.

The problem of having unfulfilled needs—or would that be a blessing?—made me think about social entrepreneur David Bussau, who'd been awarded the honor of Senior Australian of the Year around the time Bill and I started sparring over the uses of boredom. In fact, I'd just interviewed Bussau and read his biography, so the details were fresh in my mind.

A construction magnate who amassed a multimillion-dollar fortune by the time he was thirty, Bussau is the cofounder of Opportunity International, a global microfinance initiative that provides small business loans to individuals in developing nations. Like Dr. Muhammad Yunus, who won a Nobel Prize for similar work, Bussau believes self-determination, not charity, is the way to transform lives and local economies. But perhaps the most amazing thing about this man, who is now in his sixties, is that he has achieved all this despite growing up as an orphan—abandoned first by his father, then relinquished by his mother. In his view, however, he has achieved all this *because* he grew up as orphan. "The fact that my parents abandoned me was probably the greatest gift they could have given me," he told me.

I'd found this mind-blowing. Yet when I shared it with the kids, they were unfazed. "I can understand that," Anni said. "It'd be kind of fun to have the house all to myself."

I closed my eyes. Really, what do you say? Their only experience of "abandonment" was being left for the night with a babysitter who refused to serve them hot chocolate in bed.

Now, obviously no one would advocate leaving your kids on the doorstep of an institution. (Fantasize about, yes. Advocate, no.) And, Lord knows, Bussau's perception is an unusual one. Many children would be, and have been, crushed by similar circumstances. But the truth is, many haven't—and some have thrived.

Frankly, even acknowledging this as a possibility seems a subversive thought. The whole notion of allowing children to tough things out for themselves has disappeared from our parenting without a backward glance. And that very much includes finding a personal solution to the personal problem of boredom. "A man must assume the moral burden of his own boredom," admonished Samuel Johnson. Yet as parents, and perhaps particularly as mothers, we tend to assume the moral burden of *everyone's* boredom.

I'd been listening to my kids bleat on about being bored practically from the moment of conception. If their ultrasound photos had a caption, I have no doubt it would have been: "Muuuuum, there's nothing to dooooooooooooooooo in here!"

I often think about our first big trip to the States, when they were seven, five, and three. Just for the record, getting from Perth, Western Australia, to my sister's place on eastern Long Island takes twenty-eight solid hours of travel, twenty-four of them airborne. When traveling with children, the kid-chill factor makes it feel much, much longer. On this particular trip, we left for the airport by taxi in the wee hours of the morning, wending our way through the neighborhood until we reached the main highway—a distance of perhaps five blocks. At the red light, the baby tugged on my sleeve. "Are we in New York yet?" she lisped. "'Cause I'm bored!"

The assumption that it was my job to remedy life's boring bits (or, preferably, to prevent them) had never seriously been questioned—by any of us. I don't think that makes our family particularly unusual.

Boredom is a big issue for parents today. Not just listening to kids complain about boredom—but responding to those complaints. Taking responsibility for those complaints. And, perhaps above all, throwing technology at those complaints. Somewhere along the line, providing "stimulation" became a key aspect of our job description. The belief that a stimulated child is an advantaged child is so widely shared we rarely bother to articulate it. So too, of course, is its corollary: that a bored child is an at-risk child. In fact, the moral imperative to keep our kids occupied or suffer the consequences is one of those unexamined articles of faith that has helped to make modern parenting such a minefield of misplaced guilt and misdirected resources. (Baby Einstein, anybody?)

Even before The Experiment, I'd started to wonder whether we'd been confusing "plugging in" with "switching on"; whether boredom— far from being the enemy of all that is educational—might turn out to be our friend.

When we contemplated taking the leap of faith into screen-free living, there were many things we feared. Gaining weight. Losing friends. "Missing out" (in some vague but disquieting way). But our greatest fear of all was the one that Bill had articulated right from the git-go: that without our media, we'd be bored.

How ridiculous. Of *course* we were bored. Paradoxically, though, we found reconnecting with our inner blank slate wasn't nearly as gruesome as we'd feared, once we got the hang of it and rediscovered the lost art of staring into space. And allowing ourselves to be "in the moment" with boredom did motivate us—each in different ways—to discover ways to plug up the gaping, screen-sized holes in our imaginations.

For my part, I amused myself by turning to the study of boredom. I read most of Patricia Meyer Spacks's compulsively interesting *Boredom: The Literary History of a State of Mind* on the train (the looks I got from my fellow passengers were pretty entertaining too). Along the way, I learned that boredom is first and foremost an idea—a set of beliefs and values. Boredom is not a universal experience—like hunger, or the urge for straight bangs—but a product of culture. And a fairly recent product of culture at that. In fact, the word "boredom" did not even exist until the eighteenth century. And some historians argue that the concept of boredom, and by extension the experience of it, didn't either.

For that matter, "interesting" (in its current sense) was also an eighteenth-century innovation, making its first appearance in Laurence Sterne's *A Sentimental Journey* in 1768. "If life was never boring in pre-modern times," notes Spacks, "neither was it thrilling, interesting or exciting, in the modern sense of these words."[1] That's different from saying that people did not, by Bill's standards or yours or mine, experience these states of mind. One thinks of the sheer tedium (to us) of agricultural tasks such as hoeing or planting or harvesting— or of old-fashioned rote-learning of poetry, or Bible verses or times tables. Such activities may not have been precisely relished, but to have experienced them as "boring" implies the existence of an alternative. A better offer forgone. In the absence of such an alternative, you might feel blank or unmotivated or, as we say, "on autopilot." But when there really and truly is nothing better to do, you are unlikely to feel bored.

When my kids were babies, I found staying at home and being a "housewife"—despite the fact that I was divorced (LOL) and despite the fact that I felt I was doing "the right thing"—to be supremely boring much of the time. My mother never did. At least partly, this is because I could imagine other options.

The problem of boredom is also completely tied up with leisure, and specifically with the separation of work and leisure in our lives. Although technology is today blurring some of these boundaries— allowing us to snuggle up to our in-boxes in bed, for example, or to Twitter our way through tedious meetings—most of us still take the work/leisure divide for granted. Not everybody else in the world does, or has. Premodern people didn't. Those who today live in subsistence economies don't either. Nor, for that matter, do very small children (for whom everything is play), or new mothers (for whom everything is work), or genuinely addicted workaholics (who have forgotten how to tell the difference). As the proportion of work to non-work decreases, leisure itself becomes the "problem"—which is something most people rarely think about in relation to work/life balance, until they retire and freak out.

At the opposite end of the life course, we see the problem of leisure in the phenomenon of "the hurried child," as described by psychologist David Elkind in his classic 1981 book of the same name. Nineteen eighty-one—the very same year IBM introduced the personal computer (which, btw, retailed for $2,800 and boasted a 64K hard drive, which is enough to store about *three one-hundredths* of a single song). We all know kids who are like this: so scheduled they practically need a press secretary to keep track of their obligations and appearances. Sussy's school seems to specialize in them. "It's difficult keeping up with eleven-year-old Chloe Hetherington," enthuses a typical feature in our local paper. "Three times a week, the Cottesloe girl arrives at school by 7:30 a.m. to take part in music lessons; on Wednesday afternoons it's dance practice, Thursdays she's at debating, and by Saturday she's charging around a hockey field." The usual suspects counsel restraint—in this case, a school psychologist (whose own eight-year-old daughter "takes dancing twice a week as well as guitar lessons and circus sports") and a "parenting educator," who offers the insight that "It's all about balance." But it isn't difficult to detect the approbation behind the stock warnings. An all-dancing, all-singing, all-hockey-stick-wielding child practically screams parental success. In Chloe Hetherington's world, there's no room for staring into space. Boredom, it is clear, belongs to lesser beings.

As Chloe's example suggests, boredom is in part a class issue. (Producing a "hurried child" is as unmistakable a case of conspicuous consumption as driving the Audi to Pony Club.) Like parenting itself, boredom is a social construction. Although we commonly speak of boredom as if it were an objective, almost biological state of being, it isn't. On the contrary, it's more an explanation—or even an excuse, really—than a condition. It is also, and perhaps especially where our children are involved, a *judgment*.

As parents and educators, we increasingly fear that judgment—and we are willing to go to extraordinary lengths to avoid it. I was amused to read that in the eighteenth century, when the rise of bourgeois society allowed a measure of leisure time to the working classes, country folk desperate for entertainment organized grinning competitions—because that's exactly what I used to do with my kids on those rare occasions when we braved a non-family restaurant (basically, anywhere that didn't feature placemats with jokes). "But where are the crayons?" they'd wail, as I tried to explain the difference between a white linen tablecloth and a sketchpad. Sooner or later, I'd be sitting there in my cocktail dress and high heels, cross-eyed and with a pinkie stuck up my nose. Even as teenagers, their restaurant attention spans remain gnatlike. We place our orders, and they still ask me, "How soon will our meals arrive?" as if they believe I have prepared the pad thai myself and smuggled it into a back room while no one was looking.

Freedom from Boredom has emerged as a key corollary to the Digital Bill of Rights—and those who abridge it run the risk of provoking what Hannah Arendt called "the primitive anger of unfulfilled entitlement." An article I read at the start of The Experiment

advised teachers to "give up the struggle" to prevent children from
text-messaging one another during class, citing a University of Tasma-
nia study dubiously titled "2 text yrm8 is gr8!" The study found that
more than 90 percent of ninth- and tenth-graders—including those in
schools with strict (LOL) no-phone policies, regularly engaged in the
practice. Author Martin Beattie urged teachers to abandon their fortifi-
cations and start incorporating messaging into school routines instead.[2]

Nevertheless, as Bill correctly surmised, boredom—far from being
an energy-sucking black hole to be avoided as assiduously as a Mor-
mon door-knocker at dinnertime—has actually served to fuel human
progress, and many experts have noted as much. Bertrand Russell was
one of them. Philosopher, logician, mathematician, historian, social
reformist, pacifist, Nobel Prize–winning author, and serial monoga-
mist—what, no field hockey?—Russell believed boredom to be "one
of the great motive powers throughout the historical epoch." He
clearly knew whereof he spoke. But Russell's comment also suggests
that a world *without* boredom would be dull indeed—and this was a
paradox I found myself revisiting continually. "All endeavor of every
kind," Spacks reminds us, "takes place in the context of boredom
impending or boredom repudiated."[3]

Other commentators, I discovered, have seen boredom as a character flaw, a social disease, a form of passive aggression, and even as
an excuse for active aggression, arguing that people shoplift, or binge
drink, or shoot others not because they're "bad" but because they're
"bored." As a result, many of us are not simply averse to boredom, we
are frightened by it.

As someone who literally reads the fine print on the conditioner
bottle while in the shower, I found I could relate. Later, when I
was able to Google it, I discovered there is a name for this disorder:
thaasophobia—fear of boredom. Pronounceable or not, I believe it
has reached epidemic proportions in our culture. Pre-Experiment, it
certainly had done so in our family.

I expected the Digital Natives to grow restless without their media. But my own hyperelevated need to be . . . well, not "entertained" exactly, but *distracted*, was something I'd failed to factor in. After all, I was a grown-up. When I wasn't putting fingers up my nose. Like most other grown-ups, I often bragged that I was "never bored." That I—not unlike little Chloe Hetherington—was too busy to be bored. What I hadn't admitted was that I was almost always siphoning some form of input. Maybe I didn't fall asleep to Super Mario, or zone out to an endless loop of quasi-inappropriate YouTube videos, but my head-space was, in its own way, as colonized by content as anybody else's.

For starters, like many another educated adult, I consumed "news" in the same way that I consumed Coke Zero: in great empty gulpfuls throughout the day. It was filling but hard to digest, producing an uncomfortable informational flatulence. Nevertheless, I was used to taking the moral high ground and pretending to a self-evident "need to know." "Hardly a man takes a half hour's nap after dinner, but when he wakes he holds up his head and asks, 'What's the news?'" Thoreau observed a century and a half ago, with palpable disgust. What on earth would he have made of my NPR "desktop ticker" extruding headlines across my laptop screen every second of every minute of every hour of every day?

The question of what we DO with the news we "follow"—like a loyal fan, or a stalker—is one of the least addressed issues in contemporary journalism. It is something Thoreau started thinking about from the very dawn of the digital age. "We are in great haste to construct a magnetic telegraph from Maine to Texas," he wrote, "but Maine and Texas, it may be, have nothing important to communicate." Lord knows, in the age of Twitter we've stopped worrying about such minor details. Nothing important to communicate, a cynic might observe, is not only no impediment. It seems to be the whole point.

Before The Experiment, I hadn't given much thought to my own thaasophobic tendencies. Now that I have, I realize I am not necessarily typical. Not all of us feel the need to download a digital copy of *Wuthering Heights* upon hearing the announcement of a ten-minute track delay. I did exactly this the day I got my iPhone, and I'm humiliated to report that it delivered a thrill that was borderline erotic in its intensity. "As God is my witness, I'll never be bored again!" I exulted. You could practically hear the overture to *Gone With the Wind* over the hissing of the air brakes.

Don't get me wrong. I still think the App Store is among the greatest secular miracles of our age—and when I read somewhere in the second trimester of The Experiment about its billionth-download milestone, I paused for a moment of silence and longing. But I am reluctantly mindful of Thoreau's warning that "our inventions are wont to be pretty toys which distract our attention from serious things." Not that I ever really confused my virtual Zippo lighter for the torch of learning or anything, but the human capacity to be seduced and sedated by bright, shiny objects should never be underestimated. The Canarsie Indians who sold Manhattan for $24 worth of trinkets are still our spiritual brothers.

When my own trinkets were taken away, I whiled away many an abruptly empty hour considering how the flight from boredom, so-called, has been systematically impoverishing all our imaginations. I'd certainly taken the "boredom defense" at face value, accepting uncritically that my children "needed" stimulation; that without it they'd be deprived—and, by implication, potentially destructive or intrusive. Like so many other modern parents, I'd taken it as a given that even tiny babies experience boredom. I never stopped to ask myself exactly what that meant. Like, when a three-month-old watches her fist as if it were the latest episode of *Scrubs*, what on earth can boredom mean?

I recalled in high-resolution, cringe-making detail how unhesitatingly I'd diagnosed a case of premature boredom when Anni was that age and had trouble settling for naps. The interior of her crib must be too dull for her twelve-week-old sensibilities, I decided—its raucous profusion of music boxes, mobiles, activity toys, and stuffed animals ranging from teddy bears to stingrays (seriously, the kid had a stuffed stingray) notwithstanding. I was encouraged in this delusion by Penelope Leach, whose book *Your Baby and Child* was pretty much The Dummy's Guide to Motherhood of its day.

Leach's view was that "fussy" babies, as she called them, were simply understimulated babies trying to communicate a need for better programming options. She was big on DIY boredom-busters, such as mobiles of dangling tea bags and Christmas balls and teaspoons, or "whatever is to hand."

I made one out of child-safe fishing tackle that would have put Alexander Calder to shame . . . in so many ways. I staged mini–puppet shows, and worked for hours creating entertaining balloon faces. My masterwork—a yellow skull-shaped number grimacing as if from sleep deprivation—was something of a self-portrait.

Baby Anni's resolute failure to be amused by any of it suggested (to me and to Penelope, anyhow) that I simply wasn't trying hard enough. Leach hinted openly that difficult babies were probably super-intelligent. It took me some months to wake up to the fact that, regardless, she was also super-exhausted. She didn't need more entertainment. She needed less. Like mother, like daughter: She needed to sleep.

Boredom is a bit like spastic colitis. It is massively overdiagnosed. Also like spastic colitis, we forget that it is essentially an effect, not a cause. Patricia Meyer Spacks refers to the word boredom's "capacity to blur distinctions." When we say something is "boring," it is "an all-purpose term of disapproval."[4] It's not dissimilar to describing a baby's crying as "colic"—or for that matter an adult's failure to thrive as a case of "low self-esteem." Spacks, who happens to be a mother as well as a scholar, notes how often boredom is invoked as a screen for more difficult emotions within family life. She refers to "the hidden

aggression—every mother knows it—in proclamations of boredom."[5]
Boredom implies victimhood, and even a quasi-self-righteous anger
directed at the perceived source of the deprivation (i.e., you!).

An inability or unwillingness to engage may be a side effect of
physical fatigue, as we've just observed. Children who are sleep-
deprived find *everything* boring (just as their mothers and fathers do).
Less obviously, boredom may also mask *fear*: the fear of failing at some
new undertaking, for example, or within a new social setting. Boredom
can be erected almost as a shield, a force field protecting us from potential psychic harm. As the expression "numb with boredom" suggests,
it can also function as a kind of psychic anesthetic. The real source of
discomfort is blunted, or supplanted altogether—which is why, over the
long term, addressing boredom by treating it with escalating doses of
"entertainment" is a dodgy excuse for a cure. Interestingly, psychoan-
alysts have observed that boredom and clinical depression are closely
related. "How weary, stale, flat, and unprofitable seem to me all the uses
of this world!" moaned Hamlet. Translation: "Muuum . . . I'm bored!"

My own experience with boredom also suggests a connection to
loss of control. Sitting trapped in a classroom, or at the laundromat,
or on a station platform, or in a long-term relationship, or even a perfectly nice foreign country, may be labeled "boring"—but it's really
frustration borne of powerlessness. The resentment we feel at such
times may get massaged into something more passive, more socially
acceptable. Instead of getting mad, we zone out. In situations where,
on the contrary, we perceive we do have what psychologists call "locus
of control"—regardless of the level of stimulation we receive—we are
less likely to invoke boredom. Even the illusion of choice helps us to
reduce boredom's dead weight.

Paradoxically, too much choice can also induce boredom, or
at any rate indifference—almost as if an overload switch has been
tripped. An oft-cited study that found shoppers bought more jam the
fewer varieties they had to choose from is a sweet illustration of the

numbing effect that "options overload" can produce. Thirty years ago, when cable television was an innovation, the joke that you now had access to one hundred channels and there was *still* nothing on seemed the height of irony. Today it's more in the nature of a truism. The dilemma has been noted by many observers, among them Orrin Klapp in *Overload and Boredom*, who points to the "major paradox that growing leisure and affluence and mounting information and stimulation . . . lead to boredom—a deficit in the quality of life."[6]

The more interesting life becomes, in other words, the more boredom we are doomed to experience. Kinda fascinating, really.

January 19, 2009

Electricity still awesome.

Bill and Anni to see *The Curious Case of Benjamin Button* today . . . together. (The Curious Case of Socializing Siblings?!)

B. lobbied for reimbursement—"Thanks to you and YOUR experiment, we have nothing else to do!" Frankly too stunned to object. Last time they saw a film together was literally last century.

Had hoped to save money, but can see error of expectation now. Between movies—FOUR this week for me alone—books, music lessons, and CDs (we're allowed those, thank God), am cleaned out for the month already. "And don't forget you promised to reactivate our gym memberships," A. scolded. "It's the least you can do."

Okay, okay. So in the early weeks, we hadn't quite gotten the hang of "assuming the moral burden of our own boredom." I was still pretty much carrying the can for all of us—and offering cash compensations, no less, when I let the side down.

That same week, I went to a barbecue and found myself surrounded by a knot of admiring parents, avid to know how we were surviving. Honestly, I hadn't been called "brave" by so many people since the last time I took the kids to midnight mass. One man, the deputy principal of a prestigious private boys' school, told me he'd recently been ordered to smart wire the residence hall to allow the boarders "equal access" under the Digital Bill of Rights. "Parents nowadays consider Internet access an 'essential service,'" he explained bitterly. "I think it's nuts, but . . ." He shrugged. "I guess no one wants their children to feel deprived."

I smiled just a little stiffly at that.

Back at Test Pattern Central, the deprived ones were starting to find their sea legs. Within a day or two of the blackout, Bill had fished his saxophone out of the toy closet, where it had long lain abandoned like some brass Velveteen Rabbit. Listening to "Summertime," played on the deck after dinner in the waning light of a still-sultry midsummer evening, was my first moment of pure joy during The Experiment. "If it never gets better than this," I mused in my journal, "I don't care. It's already been worth it."

It had been a ridiculously long time since I'd heard Bill play anything that didn't involve a joystick or a mouse. Yet fewer than two years before, he was taking weekly lessons with a teacher he loved, and had even started talking about the possibility of a musical career. Then . . . nothing. By the end of Year 9, he'd discovered water polo, World of Warcraft, and Windows Live Messenger. MySpace, Side-Reel, and a terrifying procession of first-person shooter games followed in swift succession. Music disappeared from the horizon, as if it too had been picked off by a sniper's bullet. From time to time Bill would talk vaguely, almost nostalgically, about picking up his instrument again one day. As if music were a childish thing he'd put away along with his Meccano set and his beloved vacuum cleaner (the one he'd found on a rubbish heap and used to perform party tricks with marbles).

He'd also acquired an iPod and seemed more focused on amassing music rather than making it—or even, necessarily, listening to much of it. I'd noticed the tendency in other teenage boys. ("Do you only have eight gigs? Aw, too bad, man. Me? I've got one-sixty." Spit, swivel, and swagger, stage left.) They'd compare hard-drive capacity the way earlier generations boasted about horsepower or rifle caliber. At the same time as the iPod encouraged Bill to get excited about acquiring the ultimate playlist—mostly, it has to be admitted, by file-sharing stealth (polite terminology for breaching copyright)—it helped push music from the center to the periphery of his consciousness.

You listen to your iPod while you do other stuff, after all. That's the beauty of the device. In fact, that's the whole *point* of the device. It allows you to live your life to a set of soundtracks of your own devising (homework music, bus music, workout music, maternal-nagging-block-out music, etc.). But a soundtrack, as the name suggests, is something that plays underneath the main event. It provides atmosphere, not plot; background noise, not foreground action. It is a takeout coffee in a cardboard cup, consumed in careless sips on the way to work. It's not breakfast.

"If a man loses pace with his companions, perhaps it is because he hears a different drummer," Thoreau observed, in what is surely the best-known line in *Walden* and perhaps in all of American literature. "Let him step to the music which he hears, however measured, or far away." Like everybody else in the civilized world, I was familiar with those words. But The Experiment made me hear them anew—with unstoppered ears—as an absolutely uncanny evocation of the age of iPod. Each of us stepping to the music we hear? Hello? That is sooo not a metaphor anymore, but literally and explicitly what technology invites us to do. The fact that "pod" is derived from the Latin for "foot"—so that it could be translated roughly as "I step"—makes the connection all the spookier.

As I write these words, I experience an almost visceral longing for the personal drumbeats of my own iPod. (Put it this way: It's no coincidence that the Winter of Our Disconnect has coincided with the Winter of My Lapsed Gym Membership.) Yet I've become acutely aware that there are benefits to undergoing an elective iPodectomy. In some ways, plugging in can make it *more* difficult to hear that elusive different drummer. And as for the guy on keyboards, don't get me started . . .

I should clarify here that using our iPods (or in my case an iPhone) was verboten at all times during The Experiment, as opposed to computers, which we were able to access at school or work or friends' places or cafés, or anywhere, really, outside the boundaries of our property. (Luckily our nearest neighbors had access codes in place, or my kids might have taken up residence on the sidewalk.) "Kind of arbitrary rules," my friend Mary sniffed. But then she's a Presbyterian, so she would say that. Yes, there was a degree of flexibility, as we Anglicans say, in the way we—or I, really—interpreted the "no screens" injunction. We were allowed to listen to CDs, and the radio, of course, but any form of docking station was strictly out of bounds. It wasn't logic that dictated this decision—an audio file is an audio file, after all—it was pragmatism. If iPods were allowed to roam free in people's bedrooms, I reasoned, sooner or later they would be sure to migrate to people's ears, and lodge there like mites. Allowing iPods into the equation would have been like, oh, I don't know, decking the halls with bowls of Lindt truffles while you were on Atkins (which Sussy and I were by month four, btw). Like, why would you do that to yourself?

There was one loophole the kids did manage to slither through: their phones. Like all moral lapses, this one happened when I was looking the other way. Let me explain. Way back on Christmas Day, some time between the carving of the holiday roast salmon and the popping of our homemade crackers, Bill had suddenly shattered the merriment as if with a beribboned Christmas sledgehammer. "What about our phones?" he'd asked. I watched the merriment drain from my children's faces as, hollow-eyed and beseeching, they turned to me. The mood had lurched in an instant from Norman Rockwell to Edvard Munch. (But then holiday dinners can be like that.)

The truth was, I'd already considered the phone question in some detail. I knew that The Experiment would need to entail a total iPhone disconnect for myself. It was my chief and most cherished addiction; plus, the device itself functioned as a kind of super-screen. It did everything all the individual devices could do, and more. I could watch TV on it, or movies, or check e-mail, or surf the Net, or take photographs, or play games, or listen to music, or . . . Okay, down, girl.

The kids' phones were different. Between the three of them, they'd lost more mobiles than molars. It had gotten to the point where we all accepted there was no point getting them anything fancier than a digital tuna can and a piece of string. Although their phones did undoubtedly possess screens, they had no Internet, dodgy cameras, or crappy games. Plus, I'd resisted putting Anni and Bill on a plan and had instead made each of them responsible for buying their own prepaid minutes (an ignominy, if they were to be believed, on a par with being sent to school in crocheted Day-Glo ponchos). As a result of this, their phones functioned largely as pager devices. They could receive calls and texts, but their capacity to reply was restricted and often nonexistent. As for Sussy, she didn't have a phone at all at that point. I'd given her my brand-new Nokia when the iPhone and I were still on our honeymoon, but within a month it had been stolen by dudes unknown at a down-market Year 9 social event.

Given all that, I'd decided that the phones would be my bargaining chip—to be played if, and only if, I encountered huge resistance to the proposal. The thing was, I *hadn't* encountered huge resistance, and in retrospect I can see that that's what threw me.

Reader, I caved. But I did put some restrictions in place. No games—not even crappy ones. (Anni, otherwise the scholarly child, was especially prone to Tetris bingeing at times of stress.) No plugging in to MP3s (despite having the storage capacity of a pill dispenser, a de facto iPod was a de facto iPod, and off-limits). They could use their phones, as phones—and nothing more. And they had to agree that abusing the privilege would mean its withdrawal without further notice. "Yes, but who's to say what's 'abuse'?" Anni began.

"Do you want your bribe, or don't you?" Sussy hissed. I couldn't have put it better myself.

With access to unlimited downloads on a "real" phone—an iPhone, BlackBerry, or other smartphone—The Experiment would have been severely compromised, if not entirely pointless. Increasingly, and especially for teenagers, mobile technology is where it's at, and content provision for handheld devices is widely regarded as the next big digital frontier. Games, movies, social media, live broadcasts, the full range of rich media right there in every child's back pocket is already a reality for many. Soon it will be for most. What this will spell for our children's collective experience of "boredom" is anybody's guess. (A generation ago, we'd grown accustomed to a commercial break every seven minutes. Today's Twitterati grow restless after 140 *characters*.)

What is certain is that we will pay dearly for the privilege in all kinds of ways. According to one recent UK study, today's teenagers are spending twelve times more than they did in 1975 to buy, quote-unquote, essential technology, even indexed for inflation.[7]

The upper limit on even traditional modes of mobile connectivity, such as texting, seems to have no upper limit. "Can you believe that in April 2008 one teen managed to rack up over six thousand text messages in one month?" asks a high school counselor/ blogger. Two hundred texts a day? Ha! Not only can I believe it, I've streamed it live in the discomfort of my own home, but more of that later.

At least prepaid phone minutes are self-limiting; therein lies their advantage. Once the allowance is gone, it's gone—more or less. iPods, on the other hand, are not in the slightest degree self-limiting: While purchasing content from the iTunes Store can be fun, it is certainly not necessary. That anytime/anywhere quality lies at the heart of the iPod mystique, and was exactly what made me nervous about them. As the mother of three teenagers, I knew the first law of digital dynamics like the PalmPilot of my hand: That which can be accessed *will* be accessed. I'd fallen under the spell of my iPod from the first too—the date was January 30, 2007, and the song was the Pointer Sisters' "Jump" (and I did—with joy). Suffice it to say, I knew firsthand how the seduction worked.

Naturally, the kids agitated to be allowed to use their iPods in the car. "You said no screens in the *house*," they reminded me. "You didn't say anything about the car."

"Ah, but the car is an extension of the house," I extemporized glibly. Honestly. You needed to be a cross between Yoda and Sonia Sotomayor just to survive a day with these people. The truth is, I enjoyed listening to podded content through our iTrip—the doohickey that plugs into the cigarette lighter at one end and your iPod at the other to produce sound, improbably, through your car radio— as much as anybody. What I hated was the endless negotiations, the disc-jockeying for position, that went along with it. Whose iPod? Which playlist? For how long and at what volume? And what about podcasts? Sheesh. It almost made me nostalgic for the old days, when the only thing we had to fistfight over was the radio.

One of our old cars—a junk heap of a Mercedes wagon with

balding velour upholstery and a wheezy pneumatic locking system—had a vintage Telefunken radio that only received one station,
Fremantle Community Radio (think "Hits of the War Years" and
nightmarish Pat Boone tributes). Basically, we had two choices: We
could listen to Vera Lynn singing "The White Cliffs of Dover," again,
or we could sit in silence. It was horrible but also wonderful. Restful,
even. A balm to the decision fatigue that, as a single mother, was a
constant occupational hazard.

When our next car turned out to have both a functioning radio
and a cassette player—call it a splurge, okay?—the girls tried their
hand at creating mix tapes. These were, to use their favorite word,
"random." One of the classics featured "Sixteen Going on Seventeen"
from *The Sound of Music*, "Teenage Dirtbag," and an indie hit with
which we could all, from time to time, identify: "I Love You 'Cause
I Have To." There were several such tapes and we played them over
and over and over again, until they were threadbare and raspy. But we
loved them—I as much as anybody. In time we all learned every word
to every lyric and would shamelessly sing along. Sure, I was partial to
the Disney hits (come on: Sebastian the lobster? Does Western music
really get any better than that?) and gritted my teeth through the literary wilderness that is a Hilary Duff lyric, but rolling with the musical
punches was the whole point of our mix tapes.

Then we got our iPods, and we didn't have to anymore. For us, as
for most families, the advent of the personal music player ushered in
a whole new era in family car trips. Boredom, once as much a feature
of the average car journey as baiting your sister in the backseat, was
banished like a naughty child. Now everybody could listen to their
own content, podded in their own private world of sound. "Be your
own telegraph," Thoreau had exhorted his readers back in 1848. A
century and a half on, we'd done better than that. We'd become our
own transistor radios.

Three kids, three iPods, three playlists, and a single pervasive silence. Presto. Of course, if you listened carefully you could hear their preferred hits leaking out of their respective inner ears—from Miley Cyrus (Suss) to Metric (Anni) to Rage Against the Machine (Bill, ironically). No conversation was necessary. Indeed, no conversation was possible. As for accommodating somebody else's musical taste, well, if you didn't have to, why would you? "We know what we like, Mum," the baby informed me as she cranked up the jam on "Sexy Back." I prayed she was mistaken. I was pretty sure it was more accurate to say she liked what she knew. And as long as we stayed securely corked within our own musical marinades—never bored but arguably never stirred, or shaken either—how could we know anything else? Listening to their music had certainly expanded *my* horizons in the pre-Pod era. Few women of my age could rap as effortlessly to "Fergalicious" as I.

But it wasn't just loss of standards that worried me. It was loss of hearing. According to recent studies by the American Academy of Audiology, more than one in eight kids suffers from irreversible, noise-induced hearing loss—and the trend, much like the volume control on our kids' iPods, is definitely on the upward swing. By the time the iGeneration reach their sixties, it is estimated that more than half will be hearing impaired. Forget about earbuds. By then, it'll be a case of ear trumpets. The problem is that heavy exposure to sounds delivered directly to the inner ear through earbuds or headphones wreaks havoc on the delicate hair cells of the cochlea. How loud is too loud? Basically, if you can hear even the faintest strains of *High School Musical 3* from your child's iPod, she is listening at unsafe levels.

But trying to get my kids to listen to this—or for that matter to anything else—was unheard of. Literally. "Twenty-five percent of all iPod users listen at levels loud enough and long enough to cause damage," I found myself practically screaming one morning.

"What?" they replied.

A few days before we left for Gracetown, a fourteen-year-old British girl stepped out in front of a car as she listened to her iPod, and was killed. A witness told police, "She did not hesitate, she did not stop, she did not slow down or look before crossing." The driver was allegedly speeding and was charged with careless driving. But still. A UK insurance company claims that "podestrians," as the Brits call them, now account for one in ten minor road accidents—and over half of them involve kids. I try hard not to be overly precious about my children's safety—and they are, on the whole, very aware and responsible kids—but this terrified me.

iPod muggings were another concern—especially reports of teens so invested in their devices they were unwilling to give them up without a struggle. "The iPod to the teenager is the baby to the mother," a child expert told *The Canadian Press* in November 2008. "These kids are so invested in their music and in their playlists, it's like they put their identity into their song selection. So you're not just stealing a device. To them, you're stealing an identity."[8] I found that perfectly understandable, which is perhaps why it rattled me. I was sure I would have reacted in exactly the same way at that age. If someone had attempted to mug me while I was lugging my entire collection of LPs plus my turntable and speakers around the city with me, I too would have resisted. (As if life without Don McLean could possibly have meaning!) But I think the point is pretty plain: I wouldn't have been doing it in the first place.

For all these reasons, I decided that going pod-free—as free as the grass grows, as free as the wiiiiind blows, pod-free to follow your heart—was the way to go. If boredom was the price we had to pay for rediscovering our original headspace of environmental adaptedness, aka the lost art of staring into space, then so be it.

Not surprisingly, I guess, my own reentry trauma was relatively short-lived (but dramatic while it lasted, as you'll find out in the next chapter). As a Baby Boomer, I'd grown up in boredom's halcyon days, when all the world was black and white. The expectation that life would offer me a constant stream of entertainment was more of an acquired characteristic—and relatively recently acquired. I think back to my teenage bedroom, in the days when being sent to your room without supper still had vaguely negative connotations. By contrast, Sussy asked for—and received—a bar fridge for her bedroom at the age of ten. Between their so-called workstations, televisions, DVD players, Nintendos, and various listening devices, children's rooms today feature more entertainment options than a Las Vegas casino. In my day, if you wanted to play violent interactive games, watch inappropriate content, and converse with dodgy strangers, you had to wait for a family reunion.

But then, when I grew up, an electric carving knife qualified as high-end technology. ("Cut it on the bias, Greg! On the bias!" my mother would carp, as if the rump steak were a homemade garment Dad was attempting to piece from a pattern.) Boredom was built into the fabric of a child's life in those days. You didn't have to like it. But you were expected to endure it, and endure it you did. In church, for example—where my sister and I developed a whole host of elaborate and unholy strategies for passing the time, using props such as a single white glove, a couple of dimes for the collection plate, and a "children's missal" with the freakiest illustrations this side of Sendak. (In my mind's eye, I can still see Satan's face, green and warty like a dill pickle, as he tempted Christ on the mountain.) I can't imagine what would happen if I dragged my kids into a Latin mass and left them there for forty-five minutes, with only a rosary for a handheld game. They'd probably call the Kids' Help Line and have me arrested on charges of aggravated thaasophobia.

By contemporary standards, even our family holidays were boring.

Generally, we'd drive for many hours, with only billboards, AM radio, and clouds of secondhand smoke from the front seat to beguile the time. Parents certainly didn't take on the role of in-flight social directors–slash–events organizers for their young passengers, as we do now. A desultory game of Twenty Questions (nineteen of them some variation on "Are we there yet?") and a roll of lifesavers was as ambitious as it ever got. Lord knows, there were no Game Boys or portable DVD players, or even Wiggles CDs. The closest we got to kids' music was Johnny Mathis singing "Nature Boy" just a tad too longingly.

Our capacity to do nothing at all, and do it well, amazes me now. Even when we got to our destination—my grandmother's condo in Florida, or a family-style "lodge" in the Poconos (what were we, woodchucks?)—there wasn't a great deal of what parents today think of as "stimulation." My sister and I would loll around the pool, basting ourselves in baby oil and dreaming about being old enough to order cocktails and smoke cigarettes. For exercise, we bickered or rode the elevator. For entertainment, we read magazines or the slightly moldy books that in those days were still provided in a guests' library. (*Marjorie Morningstar* in the Readers' Digest condensed version, or—arguably more fitting for a family holiday with teenagers— a chlorine-scented paperback edition of *The Boston Strangler.*) Doing pretty much nothing at all, but doing it in a somewhat ritzier setting than our bedrooms at home, was the whole point of a holiday, as my mother would have been the first to remind us. As for "stimulation," that's what you needed a holiday *from*.

Today, the family tourism industry—and the mums and dads who keep it running in high gear—takes an entirely different tack. "Recognizing that the fickle moods of a teenager can make or break a family vacation, a growing number of resorts are spending hundreds of thousands of dollars creating elaborate hangouts to keep the adolescent set content," reported *The New York Times* in the pre-recessionary spring of 2008.[9] Most of that money is being spent on media upgrades. Like the "iChair" added by Loews Coronado Bay Resort and Spa, a kind of recliner-cum-stereo—a docking chair, if you will—"that kids can plug their iPod or MP3 player into and rock out."[10] Or the Fun Club at Cancun's Occidental Grand Xcaret, where children can play virtual tennis, golf, and bowling on a 110-inch TV—presumably without imperiling their spray-on tans. Or the no-parents-allowed Teen Lounge at the Palm Beach Ritz-Carlton, where, according to the hotel's website, boys and girls "can create their own DJ mix and upload it to their iPod, play video games, surf the Internet, play billiards, or just hang out and play pool or Guitar Hero." The Parker Meridien in New York projects Nintendo Wii games, complete with surround sound, on a twenty-foot wall located—believe it or not—on its racquetball court.

Resorts "acknowledge the incongruity in teenagers playing video games their entire vacation," the *Times* reported, "but say many parents still feel it's better than having them holed up in their room watching TV or moping around the pool."[11] Fair enough on the TV thing—but as something of an elite poolside moper myself, I resent that last crack. Yet how interesting-slash-horrifying that the resorts are taking their cue from parents on this one, that they're piling on the teen technology in an effort to assuage *our* anxiety about our kids' engagement, or lack thereof. In a sense, parenting in an age of affluence means we're all resort operators now. And then we wonder why our children behave like querulous guests with a silver-plated sense of entitlement.

Taking away my kids' room-service menu of media on demand was, in this sense, part of a larger project. And as the weeks went by, I began to see how a subtle "You can make my room up now" mentality infused our lives in other ways too. Like, why *was* I still changing their sheets and doing their ironing and generally leaving the moral equivalent of a chocolate on their pillow each night? It's not always clear where a mother's responsibility lies in these matters. But stripping away the technology smokescreen helped me to see more clearly how their learned helplessness was something I'd unwittingly encouraged. I think a lot of us do. We lay on the amenities partly out of guilt—especially if we work full-time or in some other way devote a big share of our energy to extraparental passions (whether professional, political, creative, or community-driven)—and partly out of affluence (i.e., simply because we can).

Being a single mum raised my own guilt-o-meter by a factor of twelve. I'd been dimly aware of that for years. Trying to "make it up to them" for failing to provide the expected features of four-star family life—to wit, a live-in father—had been part of my parenting agenda since . . . well, forever. The Experiment marked my declaration of independence from that doomed strategy. Giving them stuff, particularly stuff with wires and microchips, was never going to compensate for the loss of a traditional nuclear family. And if relinquishing that stuff caused our family to unravel, honestly, how closely knit could we have been to start with? It's funny, because I would never have allowed them to gorge themselves on sweets or fatty foods. But when they were in danger of becoming comfort eaters of entertainment—gluttons for gaming or instant messaging or MySpacing—I'd contrived to look the other way.

Once the Teen Lounge (home edition) had been dismantled and I found myself staring straight ahead—albeit at a blank screen—it was so obvious, suddenly, that our entire house had been set up to accommodate separate spheres. To use the language of realtor speak, I had my "retreat" (in the form of a bedroom—complete with perfectly made bed—bathroom, sitting room, and study) and the kids had their "retreat" (a separate wing comprising bedrooms, bathroom, and a "family" room)—and the twain met only in the demilitarized zone we called the kitchen. Without our personal media to structure our migration patterns, where would we *go*? Most terrifying of all: What would we do once we got there?

"They sentenced me to twenty years of boredom!" growls Leonard Cohen in "First We Take Manhattan." I played that song a lot in the southern hemisphere summer of 2009. And every time I did, I focused less and less on my yearning for technology—and more and more on my yearning for home. Maybe I shouldn't have been so surprised by that. Experts say most addictions are symptoms, not causes—ways to anaesthetize an older, deeper, more virulent ill. Clearly, my own media codependencies were no exception.

My diagnostic hunch—that I'd been using technology to treat my homesickness—grew stronger with the passage of each tech-free day. Yet in some ways, it was simply a confirmation of something I'd known all along. After all, the irony of having moved from New York, by general consensus the most exciting city on earth, to Perth, the most isolated and arguably the dullest, had not been lost on me all these years. Indeed several times I'd actually been asked to reflect on it publicly, especially in the wake of a local furor over a tourism website that had christened Perth "Dullsville." The annual Perth Festival of the Arts had even organized a public debate on the topic in which I'd participated, arguing—as you might imagine—in the affirmative. (We won.) Then, in May 2009, I was invited to address a consortium of Australian and international urban planners on more or less the same theme. The suggested topic? "Perth: An Outsider's Perspective."

Keep in mind that at this point, I'd been in continuous residence for more than two decades—giving birth to three children, buying and selling half a dozen properties, writing an award-winning volume of Western Australian history, and marrying or merging households with a dizzying array of local residents (okay, three). If I was an outsider in Perth, then Jonah and the Whale lived in adjacent suburbs.

But it was hard to work up a good head of righteous indignation. Because the truth was, I thought of myself as an outsider too. I was frank about that in my talk, possibly inappropriately so. I spoke of having felt stranded right from the git-go, and how I struggled to put down roots, like an introduced species badly adapted to its environment. I acknowledged that there were many reasons why this was so, most of them to do with who I was, rather than with what Perth was (which I knew sounded a lot like a cheesy breakup line, but yeah). I added that the lack of a critical mass of inquiring minds—given the size of our population and its geographical annexation by desert and sea—meant that our little puddle was a comfortable but stagnant one.

My audience listened intently, or at least politely. I was relieved to note how many of them were, like me, "not from around here" anyway. But afterward someone asked me a question that stopped me in my tracks. "Have you ever considered whether Perth's 'dullness' may have inspired you rather than inhibited you?" she asked. "I mean, maybe it was the *lack* of stimulation that made you so productive and sort of . . . determined." She trailed off, looking a bit self-conscious. "Do you know what I mean?"

I just stood there. Blinking like a cell phone set to silent.

January 24

Who are these people, and what are they doing in my bedroom?

9:00 a.m. Reading Saturday papers in bed as per usual. Knock on door. Anni: "Can I come in?" Grabs magazine—also as per usual, as cannot begin weekend without consulting her Mystic Medusa horoscope. Fair enough. Is a Libran, after all.

9:10 a.m. Knock on door. Sussy (in boxers and "NOFRIENDO" T-shirt): "Yo." Points to unoccupied half of king-size bed, as if to ask, "Is this seat taken?" "Go for it," I say. She snatches sales catalogues and dives in.

9:15 a.m. Scratch on door. Hazel the handheld kitten wants in. Levitates self onto bed. Notices best pillow and stakes claim. Rupert gazing up mournfully from rug. Emits snort.

9:30 a.m. Smack on door, possibly kick. Bill: "Double-u tee eff, Mum. Why is everybody in bed with you?" Me: "Dunno." Bill (contemptuously): "Losers!" We ignore him. Hazel blinks. Rupert yawns. We ignore him some more. "Well, is there room for me or not?"

February 2

B.'s first day Year 11. (Verdict: "Sucks less than ten.") Says will aim for practice goal of three hours a day. A. snorts. Rupie too (but then he would). Fought impulse to remind him first law of goal-setting is "be realistic." After all, "A man's reach should exceed his grasp or what's a heaven for?" (Always preferred Browning to Drucker.)

February 3

Anni darkened door of kitchen twice in twenty-four hours. What's cooking?! She made beautiful banana muffins this a.m. . . . entire meal for Mary and family last night. Coconut fish curry. Cucumber salad. Fudge (aka the fifth food group).

Then announced intention to compile personal cookbook. Proceeded to locate pen and paper and take dictation. Surprisingly tidy handwriting! (Although have probs not seen it since fifth grade.)

Took down recipes for spaghetti bolognese, muffins, pancakes, chili con carne, and potato salad.

Lots of laughs reminiscing re: family dinners gone wrong—"dog food" mistaken identity meal, plastic cheese incident, etc.—but truth is, was very touched. Realize that I learned to cook family fare by osmosis—mooching around kitchen watching my mother, grand-mothers. (Except muffin recipe, copied ounce for ounce from seventh grade Home Ec. and still possibly most useful thing ever learned at school.)

S. has done a bit of that, but the others? . . . not really. Suspect the cooks of the world are the kids who cared most about licking the beaters.

February 4

Wrote column in longhand again, then to X-Wray Café (nearest and cheapest Wi-Fi option) with laptop to type and send. Stressful, as have no idea of word count and tend to do twice as much as needed then scrape painfully back. Also have had to read proofs as hardcopy (sent to Mary's e-mail at work, who prints and delivers them), and call in changes. Weird, but works okay, I guess.

Arrived home just before B.—who spookily enough (given yester-day's entry) watched and chatted while I cooked dinner. (Couscous with chickpeas, sweet potatoes, raisins, and spices: bit gloppy really, but welcome break from nonstop bbq—still pushing 40 degs at 8:00 p.m., btw.)

READ ME HIS ENTIRE ENGLISH SYLLABUS.

Does that sound normal to you?

February 10

A. and self resuscitated library cards today. Even the Miss Trunchbull-ish librarian laughed at the term "blacklisted" . . . which is what we were, thanks to repeated failure to return—oh, how I winced—*The Total Makeover Book* and *How to Be Lucky.* ("Intellectual much?" as S. would say.) A. borrowed veritable treasure-trove of trash. A bit disappointing for a kid who read—and totally got—Jane Austen in middle school . . . but, hey. A book is a book is a book, right?

February 15

The prodigal returneth and surrendereth her devices. Fatted calves being scarce, we settled for homemade macaroni and cheese. Fell helplessly asleep, one arm flung over her Ducky, as of old, at 7:30 p.m.

Interesting to hear A. & B. assure S. screen-free life a "breeze" . . . compared with no lights and no power!

Big showdown between A. & B. yesterday re: possession of stereo. Had to happen. It *is* A.'s, though B. argued hard for possession being nine-tenths of law. Silently climbed to attic to retrieve old CD turntable and dismantled own amp for B.

Pretty crap sound though, so may have to spring for secondhand system, esp. now music is no longer audible wallpaper.

February 28

A. made lasagna. Excellent in that special way that only lasagna one has not prepared oneself can be. Has also been doing Sudoku and word puzzles in paper like an old retired guy on a park bench. Too cute! Also, took Rupie for two walks to beach this week. When asked why— as is totally out of character—replied, "Dunno. He just looked kinda depressed." Didn't remind her he's a pug and he always looks that way.

Both went to Murdoch Univ. to do A.'s enrollment but were told enrollment is now online or nothing. Further irony: Their system was down.

B. playing "Autumn Leaves" à la Adderley (my CD but happy to donate to good cause). Also now teaching self piano with old Suzuki books. Almost surreal to watch him in battered board shorts and Led Zeppelin T-shirt playing minuets, and picking it up like lint. To Pat's house last night and back an hour later. ("I've had my Internet fix.")

S. dividing weekend time between sleep and landline. Mostly the former. Is bad-tempered when forced to get up or hang up. Is clearly trying to hibernate way through The Experiment like some prickly teenage echidna. Possibly not a bad idea.

March 3

A. finished Gladwell's *Outliers*. (What? Nonfiction that isn't about feng shui?!) Also observed venturing beyond horoscopes and word games to actual "paper" part of paper. "I *am* a journalism major, Mum," she sniffed, as she turned to the celebrity gossip page.

B. and I fought about math tutoring today. He wants *more* of it. Random! Has been doing geometry homework at kitchen table like Abe Lincoln or John Boy Walton or somebody.

Is reading *Kafka on the Shore* (my somewhat desperate suggestion— he likes Japanese stuff so figured was worth a shot). Verdict: "Awesome." Try to hide my shock. J. K. Rowling to Haruki Murakami?! Okay, I give up. Where's the hidden camera?

Turns out Murakami's full of jazz references. Hadn't remembered that. What hooked B. was mention of Coltrane's "My Favorite Things," WHICH HE WAS LISTENING TO AT THE TIME.

March 15

Cooked roast chicken, mashed potatoes, and cucumber salad for self, B., A., and Millie. Lingered long time, picking over carcass, conversation. All of a sudden, kitchen has gone from Transit Lounge B to Command Central.

Yesterday B. described as, quote, the best Saturday of my life. Jammed with newfound musician pals—one of whom drives (OMG)—drank bubble tea x 2 and went to the beach x 2. Topped it off with a sleepover at Oscar's (evidently featuring tearful reunion with a PSP). Fascinating because it was ordinary, really—but clearly intensely so.

Later A. & S. discovered in A.'s room, side by side under covers, singing along to top-forty radio in euphoric, trancelike state. (Taylor Swift: "Love Story.") Moral: If you can't get the ringtone, *be* the ringtone.

My iPhone/Myself: Notes from a Digital Fugitive

The mass of men lead lives of quiet desperation. . . . A stereotyped but unconscious despair is concealed even under what are called the games and amusements of mankind. There is no play in them, for this comes after work.

—WALDEN, *chapter 1*

I get up at 4:30 every morning. I like the quiet time. It's a time I can recharge my batteries a bit. I exercise and I clear my head and I catch up on the world. I read papers. I look at e-mail. I surf the Web. I watch a little TV, all at the same time. I call it my quiet time. . . . I love gadgets. I'm an iPhone guy.

—ROBERT IGER, CEO, *Walt Disney Company*[1]

Like most illicit affairs, this one had started innocently enough. We were just friends at the beginning. Work colleagues, really. But you know how it is. You start having lunch together. You meet up on the train. You go for walks. And the next thing you know, you're practically living in each other's pockets. Or one of you is, anyhow.

I was always going to fall for my iPhone. I can see that now. My craving for information—a black hole of lust and neediness, incapable of satisfaction—was something I'd struggled with all my life. I'd go to the dentist and bring two *New Yorkers*, a novel, a portable radio, and a rhyming dictionary . . . just in case. Boarding an overseas-bound plane, it never once crossed my mind to fear a crash. But the possibility of settling in for a fourteen-hour flight and discovering I'd left my novel in the airport (a living nightmare that happened to me on a Qantas flight from L.A. to Sydney in 1998, and which still gives me flashbacks), now *that* could trigger a panic attack.

When the iPhone came along—which it did in Australia in mid-2008—I had been tapping my toes and hyperventilating for twelve long months. I'd been in New York in July the previous year, right after the U.S. launch, and a handsome stranger in a hotel lobby—seeing the gleam of yearning in my eye—gave me a quick induction. I loved everything about it. The way it felt in my hand, so sleek yet substantial. The way it moved, in such sinuous spins and slides. The way it responded when touched in that special way.

We'd only been together for six months, but in that time we'd developed a relationship that was totally in sync, in a totally out-of-sync kind of way. Needless to say, The Experiment shattered all that. Talk about a toxic breakup. This one had all the elements: anger, denial, bargaining. A massively overdue bill. I write these words five months, two weeks, four days, ten hours, and nine minutes—no make that ten—since that fateful day on which I told my iPhone "We need to take a break . . ." And although I would never have believed it at the time, in the end I've found acceptance.

There was never a question in my mind that getting clean after six deliriously dopamine-fueled months on the iPhone would be my own biggest challenge. But then, as far as I was concerned, most of the other screens in our house had been a turnoff for a long time.

Even the loss of my laptop, while acutely felt, was something I was able to put into perspective in the early weeks. Mostly I associated the laptop with drudgery: churning out copy and column inches to the implacable, circadian-like rhythm of daily deadlines. The prospect of writing anything longer than a grocery list in longhand (and even those I'd been known to type, format, and print) was horrifying. But I'd organized a nice long hiatus for myself. One puny little five-hundred-word column a week was all I'd have to worry about, and if I really hit the wall, I knew I could take myself off to a café somewhere for a latte and a bracing shot of Wi-Fi.

I loved Della—yes, my laptop had a name (shaaaame!)—but I also recognized that a trial separation was probably the best thing that could happen to us at this stage in our relationship.

Looking back, I can see that Della was the spouse and help-meet—faithful, reliable, comfortable, and just a little dull. But my iPhone, iNez? Hoooo, mama! Now that was one smokin' hot affair. In one impossibly sexy handful, iNez embodied all the things I love best about technology. It doesn't get any less humiliating when I stop to think about exactly what those things are. Basically, iNez was compliant, discreet, entertaining, ridiculously receptive, and looked amazing in black lacquer. She might as well have been a freaking geisha. If I'm honest—and it's killing me to admit this to myself, let alone to you—I got a buzz from being seen with iNez. I loved what she could do, but I also loved what she stood for—some heady confluence of youth and wealth and mastery. Being seen in her company made me feel important, powerful, "in the loop." But loops have a way of tightening gradually. That's why I had to leave her.

When Ian Schafer's wife complained that he would pay more attention to her if she were digital, the thirty-three-year-old CEO of online marketing firm Deep Focus didn't even try to deny it. Tweeting their exchange, he was interested to discover that "a lot of people's wives feel the same way." Schafer told *USA Today* he believes the human brain wasn't built to handle so much connectivity. "We want to do more and more," he muses, "but the more we actually do, the less of it we actually accomplish."[2]

Tony Norman, who reviews gadgets for the *Pittsburgh Post-Gazette*, was dismissive when his friends warned him not to get too close to the iPhone. He failed to lash himself to the mast and suffered the consequences. "If you ever want to know what was going through Frodo Baggins's mind as he stood clutching the evil ring over the lava pits of Mt. Doom in *The Return of the King*," wrote Norman, "buy an iPhone."[3]

Computerworld's Galen Gruman suffers from iPhone-related separation anxiety that's so severe he's developed a dread of subway tunnels. It's not the claustrophobia that gets to him. It's the "connectivity gaps." Galen confesses that at such moments he finds himself "thumbs poised . . . itching to reconnect to the outside world."[4]

Then there's Melissa Kanada, a twenty-seven-year-old PR consultant, who swore she'd stay clean during a two-week overseas trip with her boyfriend. "I think I lasted maybe four days," she recalls. "I'm like, 'He's in the shower.' Like, you feel welcomed again."[5]

And Nick Thompson of *Wired* magazine observes, "There are a lot of people who have a problematic relationship with these devices, where the device becomes the master and they become servant." Thompson would never let this happen to him. No way. Like the time he was expecting an important message but didn't want to be checking for it every ten minutes. A less imaginative user might simply turn off the phone. But with the instincts of a seasoned junkie, Nick knew more drastic measures were called for. "I took the battery out, and then I put it in the [sleeping] baby's room," where no man—no matter how desperate for his next data fix—would dare to tread.[6]

And let's not forget the guy who experiences "phantom vibrations." ("I can still feel it as if it were receiving e-mail. I reach down to check it . . . and it's not there! Am I losing it?") Or the man who dropped his phone in the toilet and fell to his knees to rescue and perform CPR on it, lest "the water-logged center of my universe" slip away to that great helpdesk in the sky. ("I tried blowing in all the holes, then I got the hairdryer out to try and salvage 'My Precious.'")[7]

I recount these testimonies not in censure, or even pity, but with the shock of recognition. They made me feel better about myself— and worse at the same time. Better, because most of these people seemed just a teensy bit sicker than I myself. But worse because they helped me to see that the problem I'd thought of as "mine"—part of my own neurotic personality structure—is actually embedding itself in our *culture*'s neurotic personality structure.

BlackBerry users got there first, of course, coining the term "Crack-Berry"—*Webster*'s word of the year in 2006—as far back as 2000, and contributing to a rich and disturbing literature on their digital drug of choice. The 2008 e-book *CrackBerry: True Tales of BlackBerry Use and Abuse* compiles the greatest hits (in every sense of the word) from the CrackBerry.com site, including some of the stories above.

"This is a self-help book on coping with BlackBerry Addiction," declare authors Gary Mazo, Martin Trautschold, and Kevin Michaluk. If this is self-help, I am Lucy in the Sky with Diamonds. Indeed, the book pretty much does for the CrackBerry what Tom Wolfe's *The Electric Kool-Aid Acid Test* did for hallucinogens.

CrackBerry is a deeply creepy read, what with featured affirmations such as, "We feel better, more complete and more whole when we are tethered to our BlackBerry at all times"—the irony being that there *is* no irony—and stories like that of "Sue," the survivor of a horrific car crash, who recalls, "I was screaming in pain and asked them to find my BlackBerry." But in some ways it gave me exactly the kind of tough love I'd been crying out for.

The "Chart of Shame," for example, which allows abusers to

locate themselves on a ladder reaching from "Plain Rude" to "Down-right Dangerous": Did I "interrupt conversation" to use my phone? (Duh.) Did I "read and respond to e-mail during a meal with others?" (Depends how you define "meal." Also "respond," "during," and "others." Oh, okay. Yes, constantly.) Did I type "while driving others in a vehicle?" (Mebbe.) Did I text "while skiing on a crowded slope?" (NO, mo-fos, as a matter of fact, I did NOT! I can't ski. So there.)

The whole cringeworthy exercise made me relive moments of iPhone passion I'd have happily mothballed along with my size 48L maternity bras, or that tiny shred of Bill's umbilicus that detached itself into his nappy and I could never quite bring myself to throw away. (I showed it to him recently—it looks not unlike a piece of snot now—because he thought I was making it up. Or perhaps he only wished I was.)

I remembered how bereft I felt on those rare occasions that I left iNez at home and had to trudge through the day alone and unplugged. I remembered the sensation of rootling feverishly through my handbag, frantic for the familiar touch of cool, tempered glass. I remembered the sick-making adrenalin surge I'd felt when I couldn't find her for a moment or two, or how, securely sealed inside my info-womb once more, I'd meet the eyes of other iPhone users on the train and we'd smile a secret smile, coyly complicit in our shared vice.

Okay, so maybe I wasn't as hardcore as some users—Johnj41, for example, who doesn't simply sleep with his phone, he *showers* with it. ("I actually keep a Ziploc bag in the bathroom just so I can do this. I know, I'm pathetic, and you know what? I don't care. LOL!") Or Lenny M., who Velcros his phone to the handlebars of his bike so he can see the screen while he rides. Or author Gary Mazo, who has a bondage-and-discipline thang going—"I can dress it up in leather if I take it out on the town and I can protect it in armor if I need to risk its being out in the world." They make iNez and me seem like junior prom dates. But a dependency is a dependency is a dependency, and the experience of going cold turkey emphasized how far in denial I'd been about my own neediness. Or maybe "wantiness" would be more accurate.

Because the iPhone, like any other smartphone, is such a heady cocktail of functionality, it took me a while to distill each of the elements. Phone, text, music, podcast, Internet, e-mail, apps. In my own way, I was addicted to all of them. But the greatest of these was e-mail. Even at the very earliest stages of my rehab—when I was still experiencing auditory hallucinations of my beloved ringtone (ironically, it occurs to me now, a stride piano rendition of "Let's Call the Whole Thing Off")—I was aware that e-mail withdrawal would be my biggest personal challenge.

I'd always been . . . well, let's say "enthusiastic" about e-mail. ("Obsessional" is such an ugly word.) And why not? In my uneasy Western Australian exile, e-mail provided a direct pipeline to the world I'd left behind, capable of eradicating the tyranny of distance at the touch of a "Send" button. As a journalist, I'd made aggressive use of e-mail right from the start. Nothing thrilled me more than getting an instantaneous response to some arcane query from a source in New York or London or Washington, D.C. The twelve-hour time difference between Perth and EST (aka "the rest of the world") meant it was easy to catch people online, checking their e-mail before bedtime, just as I was logging on to my day.

In the early years, once my workday was over—basically, whenever the kids rolled up from school or sports—so was my e-mail attention span. It never really occurred to me to go racing back to Outlook after dinner or before bedtime. My home office was nice and all, but it definitely wasn't a place I wanted to hang out in once the sun went down. Even in the more recent past, when I've had access to a laptop and wireless broadband, I wasn't much inclined to schlep Della around the house.

Once I hooked up with iNez, those relatively functional patterns dissolved. Now that I had the freedom to walk around in-boxicated all day long (and all night too if I wanted), my info-neediness went through the roof.

Going free-range, while it is very, very good if you are a chicken, is not necessarily so great in other categories of existence.

The whole thing made me think about my mother smoking in the garage. When I was in high school, my parents both decided to quit. My dad simply threw away his Kents, bought a pair of Nikes, and never looked back. My mother did things differently. She believed in preparations. For her, setting the table for Thanksgiving on Halloween was as close to "cold turkey" as she ever cared to get. My mother attacked her nicotine addiction in much the same spirit as she prepared a holiday meal: very, very gradually. In the last stages of her withdrawal, when she'd finally declared the house a smoke-free zone, she still allowed herself to smoke in two places: the backyard (where the neighbors could see) or the garage (where the dog crapped obediently on spread newspapers). It was a toss-up which environment was less appealing. I can still see her sitting on the concrete step in her winter coat, puffing on a Carlton with concentration.

Then, I just thought she was a nutter. Now I see the menthol in her madness. By tethering her habit to a specific, not entirely hospitable place—by putting clear spatial boundaries around it—she was gaining mastery over it. There was no possibility of lighting up automatically anymore. Every cigarette was a conscious decision—and a confining decision. By cooping herself up in the garage, she was segregating the act of smoking from associations with anything else, except smoking itself. There was nothing else for my mother to do in the garage *except* smoke. When she was finished, she reintegrated. She came, as it were, home.

Well, in the old days that's the way e-mail used to be. It was like that cigarette you smoked on the step in the garage. It was a habit that knew its place. In fact, you might even say that in the old days that's the way work itself was.

The fateful interface between wireless technology and miniaturization—of which the smartphone is the most recent apotheosis—has untethered work from its sense of place. Offices, desks, paper? Old media, all of them. All the world is an office now, and all the men and women merely telecommuters . . . whether we want to be or not. "It's a Pavlovian response," insists one recovering CrackBerryhead. "The bell goes off to indicate a message. I walk like Frankenstein across the room, arms out—'Must . . . check . . . messages.' "[8]

Of course, it's not that simple. Not even the App Store has figured out a way to disable our free will. Yet. Nevertheless, the weirdly hypnotic pull of the e-mail alert is among the most successful attention-seeking strategies known to humankind. Like a ringing phone, or a newborn baby's hiccuppy cri de coeur, it is very nearly unignorable.

AOL Mail's fourth annual E-mail Addiction Survey, published in July 2008, found nearly half of the four thousand e-mail users surveyed considered themselves "hooked"—up 15 percent from 2007. Fifty-one percent check their e-mail four or more times a day, and one in five do so more than ten times a day.

In the Age of the Smartphone, those figures look almost comically undercooked. (Tellingly, AOL hasn't bothered to do a survey since.) Now we don't actually "check" our e-mail at all, but just sort of inhale it continuously throughout the day.[9] We approach our messages as demand-feeders—in exactly the way I'd once been advised by a lactation counselor. "Stop thinking of breast-feeding in terms of 'feeds,' " she urged. "Nursing should be as natural and as frequent as breathing!" It sounded both lovely and terrifying. Perhaps unsurprisingly, suckling all day on one's in-box is too.

The survey did find that more than a quarter of respondents had been so overwhelmed by their in-box they'd declared "e-mail bankruptcy," or at least considered it seriously. I was fascinated to read that, because that's exactly what I'd done at the start of The Experiment. Finding out there was an actual name for it made me feel less freakish.

And then I read that "20 percent of users said they have over three hundred e-mails in their inboxes!"[10] Three hundred? With an exclamation point, no less? Who were these pussies? At the point *I* declared bankruptcy, I had 9,637 messages. Now, that's a figure worth punctuating.

I didn't close my account entirely but set up an automated out-of-office reply, effective January 4, 2009:

> I am in an e-mail-free zone till further notice.
>
> Yes, really!
>
> I am happy to receive written correspondence at
>
> 154 Edmund Street
>
> Beaconsfield, Western Australia
>
> AUSTRALIA 6162
>
> or to receive phone calls in the time-honored manner on
>
> 618 9430 4106

Clicking on "apply and save" was like going into freefall. The sheer audacity of it made my head spin. I felt defiant, reckless—as if by disconnecting, even temporarily, I were doing something illegal and perilously beyond the pale. And so I was, as responses to my (admittedly dramatically worded) announcement made clear. I'd worried about inconveniencing people, but the possibility of spooking them never occurred to me. Oops.

My family was the first to freak. "We thought you'd disappeared!" my mother cried. (Was it my imagination, or did she sound just a teensy bit disappointed?)

"We got some kind of weird bounce-back message about your e-mail being down," complained my sister in an accusing tone. Other people rang to say how sorry they were that I'd lost my job. WTF? Since when did going offline equate with being unemployed? I fumed. Then the answer occurred to me: probably about ten years ago. Considered objectively, the inference that being disconnected equaled being disenfranchised was a pretty logical leap.

Putting my home phone number out there was another of the experimental risks I felt I had to take. We'd had a silent number for many years, ever since my book *Wifework* sparked a series of abusive phone calls from ex-husbands—and not even my own ex-husbands. But it had been years since I'd been seriously troubled by hatemongers. I figured getting the odd unwanted call was a small price to pay for unyoking myself from the burden of those nine-thousand-plus messages. And, anyhow, it seemed doubtful that anybody who didn't know me well would have the chutzpah to call my home number.

Yeah, well.

The first weekend, I received a call from a reader wanting to discuss that Saturday's column. It was painful, but I survived. I even managed to sound brisk—an effect I'd been trying and failing to achieve for, oh, half a century now. I braced myself for a barrage of similar home invasions. I never got a single one.

It's interesting how few of us scruple to set up a private e-mail address, yet are willing to defend our home phone numbers to the death. For some reason, we don't regard unwanted e-mails as invasions of our privacy, or incursions on our headspace, in quite the same way. Which is weird, because in lots of ways e-mails are much *more* intrusive than home phone calls.

If you work on a computer all day—as so many of us do—e-mail messages explode in your face constantly, like tiny hand grenades hurled by an unseen enemy. Thanks to the hypervigilance of our Outlook accounts, messages from anybody, from everybody—friends, colleagues, bosses, randoms—bleat insistently for our attention all day long, elbowing their way onto the very pages of the very documents we are attempting to process. Sure, it's just a two-second flash in the corner of the screen. But how it haunts one's consciousness!*

Phone calls—even if you do happen to work from home—aren't remotely similar. For one thing, friends don't ring you in the middle of your working day to tell you a joke (let alone five friends, let alone a bad joke), or pass on a nugget of homespun philosophy, or describe an impossibly cute kitten/puppy/toddler/ferret. Retailers don't ring you with deals expressly designed for "customers like you." Colleagues don't ring you with verbatim replays of conversations they've had with other colleagues. And nobody, but *nobody*, ever rings you, says "yep," and hangs up again.

For the purposes of The Experiment, just to be on the safe side I decided to disconnect our home answering machine. I'd always kind of resented the answering machine and the way it shifts responsibility for making contact from the caller/petitioner to the receiver/petitionee. I was happy to deal with the occasional inconvenience of missing a call if it meant a holiday from playing telephone tag with people I'd likely never wanted to speak to in the first place.

Long story short: I probably did miss a lot of calls, and probably 99 percent of them were truly, madly, deeply miss-able anyway. The 1 percent that may have changed my life indelibly must remain in the category of what Donald Rumsfeld has taught us call the "unknown unknowns." I'll never know what I don't know about those calls—or even if there were any. I wouldn't say I'm exactly cool with that. I'd say I'm *euphoric* with that.

But back to our e-mail story. Heaven knows, you don't need an iPhone, or any other smartphone, to become hopelessly hooked on e-mail or to experience big-time boundary issues between work and

---

*And, in answer to your question, yes, I've tried resetting my alerts. It doesn't help. In fact, *not* knowing when messages arrive is even more distracting somehow.

home. But it helps. The (slightly sepia-toned) AOL study found that
nearly two-thirds of people check their work e-mail over a typical
weekend, with one in five doing so five times or more. Twenty-eight
percent admitted to sneaking a peak at work e-mails while on vacation
(and the remaining 72 percent were probably lying). Among respondents living in New York, a quarter wouldn't vacation anywhere they
could not access e-mail.[11] Just a couple of years on from that survey, it
all sounds so quaint. When astronauts can tweet from space, as they
did for the first time in January 2010, it's hard to imagine where on
earth—or around it—a signal-free vacation spot might be. ("Hello
Twitterverse!" wrote space station resident Timothy Creamer, aka
Astro TJ, from the vastness of the heavens. "We r now LIVE tweeting
from the International Space Station—the 1st live tweet from Space! :)
More soon, send your ?s"[12] Kinda gives you goose bumps, doesn't it?)

The AOL study, waaaay back in aught-seven, found only 15 percent of e-mail users used a mobile device to check messages. But of
those, nearly a third (30 percent) confessed that, as a result, they feel
"married to the office."

Possibly because they are sleeping with it, though not having sex
with it (LOL). Seriously, it's not just me. It turns out that fooling
around with your smartphone in bed is probably the second most
widely indulged secret shame of modern times. AOL found that 41
percent of mobile e-mail users freely admitted snuggling up to their
cell phones while they slept. Sixty-seven percent check mail in bed, 25
percent while on a date, 50 percent while driving, and 15 percent in
church. (The Lord works in mysterious ways. Why not through Outlook?) Oh, and remember the bathroom thing? Which made me feel
so shameful and perverse and wrong? Well, I'm not saying it's a nice
look to check your in-box while you're on the outbox. But learning
that 59 percent of my fellow smartphone users are doing exactly the
same thing in there definitely made me feel less unclean.[13]

In 2007, smartphones were still prestige gadgets. Only people of a
certain class got to read their mail in the john. A year later, 10 percent
of Americans owned a smartphone. By March 2010, Nielsen forecasts
were estimating that one in two would have one by Christmas—just
in time for a blazing yuletide log-in.[14]

Uncovering a whole covert world of fellow Outlook abusers made
me feel less debauched. But it also sparked an auditory hallucination
of my mother saying, "If everyone else was jumping off the Brooklyn
Bridge, would you do it too?"

Knowing that people *are* using technology in a particular way begs
the question, how *should* we be using it? Or, for that matter, shouldn't
be. Checking your e-mail while riding your bike, or showering, or
voiding your bowels, may be unsightly or extreme. But is it wrong?
Once I broke up with iNez—rather callously shutting her down and
stuffing her into an unmarked pigeon hole in my home office—I had
tons of downtime in which to contemplate that question.

Like any other smartphone—generally defined as a cell phone with
PC-like capability—the iPhone suffers proudly from "feature creep."
As well as its super-screen abilities, it incorporates an organizer, a
GPS, conversion calculators, an alarm clock, and a truly vertiginous
array of downloadable applications—as of this writing, fifty thousand
and multiplying Malthusiastically every minute—from iLickit, a
game that literally involves players licking a picture of a food item with
their real-life tongues; to FlyChat, a kind of antisocial social media
utility that sends users' messages to complete strangers, for no apparent reason; to more sober applications for doing just about everything
from online banking to detecting speed traps ("Trapster") to keeping
track of your (or a significant other's) menstrual periods ("Lady Biz")

to playing your choice of diabolically annoying sounds—think baby colic, jackhammer, slurping (the eponymous "Annoyance").

Bizarrely enough, I found being cut off from the actual phone part of the iPhone—the most prosaic and least alluring of its multitudinous charms—was the easy part. It also had the biggest impact on my life—my parenting life especially. I'd anticipated this to some degree. On a day-to-day, hour-to-hour basis, calls to or from my kids accounted for upward of 80 percent of my phone traffic. I accepted from the outset that disruptions to my parental peace of mind would be unavoidable. What I never expected was that My Big Hang Up would actually *increase* that peace of mind. My unexamined assumption that more contact must produce better parenting (and generate less anxiety) was just that: unexamined. My dependence on the phone was something that had grown up organically over the years, spreading its tentacles imperceptibly, like a nail fungus.

It was a wake-up call to find out that my most indispensable device could be so easily dispensed with. But what really pushed my buttons was realizing my 24/7 availability had actually impeded my efforts to parent well and my children's to . . . now what would you call it? "Kid" well? In some ways, the whole thing was like my discovery about the dishwasher. All these years it had felt so much more efficient than washing dishes by hand. It had looked and sounded that way too. But measured objectively in time and effort and outcome, it *so* wasn't.

The urge to stay continuously connected, like the yearning to produce whiter-than-white cuffs and collars, isn't a problem technology has solved. It's a problem technology has by and large invented. So many of our standards—of normalcy, of effectiveness, of propriety, of safety—are consequences of our technologies. This is exactly what Thoreau was getting at when he warned us against becoming "the tools of our tools." (Easy for him to say. The great transcendentalist apparently sent his dirty clothes back to Concord for laundering!)

That doesn't mean technology is evil, that it gets us to do its bidding via mind control, like some fascist dictator or bolshy preschooler. To my way of thinking, this is the technology equivalent of the Twinkie Defense. Nor does it mean that we, as the users of technology, are passive victims: hapless Pandoras wringing our hands in dismay at the chaos God hath wrought. And yet, just as the stuff we consume has a funny way of repaying the compliment, the devices meant to simplify our lives merely create new and improved complexities.

"Our life," Thoreau observed, sitting there in his nicely laundered socks, "is frittered away by detail." One wonders what a guy who thought the Post Office was frivolous would make of the iPhone.

Severing the cellular umbilicus was something I'd dreamed of guiltily for years, from the time Anni got her first phone and started essentially microblogging the events of her after-school day. "There's nothing to eat." or "Spider in bathroom!! When RU coming home to kill???" or "Bill just punched me." (I'm pretty sure she had that last one on speed-dial.) For years, text messages like these had been lobbing into my workday like softballs through plate glass. And I responded to them in a way that shocks me to remember: i.e., seriously. I'd text back earnest instructions on arachnocide, or offer snack food location tips. ("Look on the second shelf of the pantry" or "Behind the milk.") Exactly whose interests did I imagine this sort of thing was serving?

"Staying connected" is one thing—and that's what phone conversations, especially after-school phone conversations, are all about. The exchange of nuisance texts is quite another. If you'd asked me at the time, I'd have insisted that the contact was crucial, even if the content was trivial. But truth is, these exchanges did so little to maintain loving relationships or offer reassurance—and so much to encourage dependence (theirs), guilt (mine), and dissatisfaction (all around). We were "connected," all right. At the hip.

It wasn't just texting either. How many times had I turned on the phone after a meeting or a movie and found ten or twelve missed calls from home—only to ring back frantically, the adrenalin practically squirting from my ears, and hear, "Oh, never mind. We found the brownie mix/hair scissors/tape/printer cartridge/pizza wheel/front-door key/cat/toilet paper/remote/black leggings/marshmallows/deck of cards/vacuum/butter/tweezers already." I'm picturing them being duct-taped to the kitchen chairs, or huddled in an ambulance on the way to the Burn Unit, or at the very least sobbing over the lifeless body of the pug, and they're eating brownie batter and watching You-Tube. A few times I tried explaining to Anni how upsetting it is to get so many missed calls in a row, but she just shook her head sadly as if to say, "Have you considered colonic irrigation?"

I'm having lunch with a friend when her phone rings. She's placed it on the table between us—as one does nowadays—as if it were a flower arrangement, or a piece of cutlery. "My son!" she exclaims, glancing at the screen. "Sorry!" She turns slightly away to take the call, and I eavesdrop shamelessly. (What? A girl's entitled to *some* compensation for the interruptions of modern life.) The ensuing conversation is mystifying.

"Which way are you facing?" I hear her say. "No. I mean, which direction?" and then, after a long pause, "Do you see the grain silo, or the car park?" After a few more cryptic queries, she concludes, "Right. Walk up the ramp, turn left, and walk all the way down to ground level. Ground level, do you hear? You should be able to see it from there. Yeah, no. It's big. Really big. Uh huh. Love you too, honey. And if there's a problem, call me back, okay?"

She hangs up and I do my best to look uninterested. The truth is, I'd give my whole untouched half of our blue-cheese-arugula-and-pear pizza to hear the backstory. I mean, obviously the kid was lost somewhere . . . but where? Like, there just aren't that many deserted grain silos around here. Plus, Hunter isn't nine, or physically or intellectually disabled. He's twenty-two, over six feet tall, and brilliant.

Well, it turned out the kid—oh, let's be honest, the *man*—was on a train platform. Seems he'd just gotten off at an unfamiliar stop and wasn't too sure which direction to walk in. Should he go right? Should he go left? Sure, they were the only two options . . . and yes, okay, there were plenty of people milling around. But still. It was awkward.

We went back to our pizza and our conversation, but I found the whole incident pretty hard to digest. The idea that an intelligent, strapping twenty-something would sooner get his mother to act as a remote-access GPS than ask the guy standing next to him—and that neither he (apparently) nor she (as far as I could tell) thought there was anything strange about that—struck me even at the time as noteworthy. I mentally filed it under "D," for "Don't Let This Happen to You." But it wasn't until we were midway through The Experiment that I started thinking about it again. Hunter's Dilemma, I recognized with horror, was simply the logical extension of the same digital dependency I'd been fostering in my own family.

No self-respecting parent sets out to tie grown children to their wireless apron strings. But it creeps up on you. A request for ketchup here. A forwarded phone number or a little remote-control refereeing there. And before you know it, the service provider is *you*. It's like a toddler's bedtime ritual that grows just a wee bit longer each night, until you find yourself doomed to performing a sixty-minute nightly program of massage, puppetry, dramatic readings, and intercessions with powerful unseen forces. Bit by bit, your willingness to accommodate is making everybody totally helpless to live their own lives (or sleep their own sleep, as the case may be).

Soren Gordhamer, author of a quirky little self-help book called *Wisdom 2.0: Ancient Secrets for the Creative and Constantly Connected*— basically, Buddhism for people who text too much—argues that enlightened technology use is all about "seeing choice." Responding to our cell phones as if they were our masters, or our mothers—allowing their summons to interrupt our driving, our dining, our dreams— is a clear case of being blind to choice, relinquishing the privilege of choice. Gordhamer explains, "If I do not see choice, then I have none. If technology is my master, I must heed its every call, and everything else is secondary."[15] Or, in the haunting words of poet Adrienne Rich, "Only she who says / she did not choose, is the loser in the end."

I agree with all of that, yet I know firsthand how difficult resistance can be. (The Experiment doesn't prove I am good at "seeing choice." On the contrary. I fear it shows I am so bad at it I have to resort to desperate measures.) Ignoring a ringing phone—in this case, quite literally—goes against all our instincts. And that goes double when the ringtone is blaring Justin Bieber. But instincts are not reflexes. Instincts can be informed, and reformed. They can also be managed. "Human nature," as Katharine Hepburn reminds us in *The African Queen*, "is what we were put on this earth to rise above."

One of the best ways we can do that is by testing our privately held assumptions against the evidence. There hasn't been a ton of research into the impact of cell phones on interpersonal behavior, but the studies that do exist are pretty clear that cell phones do not make us more switched on. On the contrary, they seem to encourage users to be *less* socially responsible. A study conducted by Intel found that one in five people surveyed admitted to being more careless about punctuality because they knew they could reschedule via text at the last minute. One in five? I think we *all* do this.

I certainly used to. And my kids—all of them—did it constantly. Not just with me, but with their friends and one another. They were perfectly comfortable with a concept of time that was amorphous and stretchable, like the Dreamtime or mozzarella cheese. It drove me insane.

"I'll text you later!" Sussy would call over her shoulder when I dropped her off at the mall or a friend's place.

"NO YOU WON'T!" I'd bellow. "I'll meet you here at three thirty sharp!" Generally, though, I'd get a text at 3:25, wheedling for an extra half-hour, or suggesting I "chill" at a café for a while. I was so bad at "seeing choice," I'd actually text back. "If I wanted to 'chill' I would have stayed childless in the first place, young lady!" I'd snap. Or, more tersely: "Forget it. Just be there." I won these battles, but the fact that we were having them at all showed I was definitely losing the war. Each such exchange cost both of us time, money, and aggravation. It accomplished nothing, and created nothing—unless you count the ill will.

Until it stopped, I wasn't aware how much of a drag all this mobile micro-tweaking really was—and how far it *eroded* my peace of mind. And possibly theirs too. Being constantly connected meant keeping one's options constantly open. SMS made everything negotiable: hour by hour, minute by minute. There was no such thing as a firm plan or a final schedule.

There was no 5:00 p.m. sharp. Just vague intentions—a whirlwind of possibilities out of which, eventually, if you were lucky, a mutually agreeable course of action might emerge.

As far as I could make out, Anni, Bill, and Suss were genuinely comfortable with this. They've never really known what it's like *not* to improvise their lives moment by moment. Clearly, it's a Digital Native thing. For a Digital Immigrant like me, however, the psychic costs of staying chilled far outfroze the occasional benefits. Making a final decision, even about something as trivial as a pick-up time, means crossing an item off the to-do list. Keeping decisions provisional means you never cross anything off. Every day is a "work

in progress." That's creative. And it's exhausting. I loved how cel-
ling out hung a figurative DO NOT DISTURB sign on the doorknob of
my life.

MIT's Lemelson Invention Index ranks the cell phone as "our
most hated modern tool," so I am obviously in good company, at
least among my age peers. Because I'm pretty sure Digital Natives
would rate their phones as their most dearly beloved modern tool.
A University of Grenada, Spain, study of eighteen- to twenty-five-
year-olds found as many of 40 percent admitted to being "deeply
upset and sad" at missing calls and suffering "anxiety, irritability, sleep
disorders, or sleeplessness" and even "shivering and digestive problems" when their phones were switched off.[16]

Digital Immigrants tend to have a less intense relationship with
our phones. Yet we do tell ourselves—I certainly told myself—that
we need to be on constant standby "just in case." The truth is, if, God
forbid, there's a real emergency, we'll be found—just as the reception-
ist at Sussy's school prophesied.

Case in point: the Saturday night Anni needed to speak to me
urgently about her plans for the evening. I was home, but our landline had been engaged for more than an hour. Surmising, correctly,
that it was Sussy talking to Maddi on the latter's Sidekick, she texted
Sussy's friend Andy in London asking if he would please IM Maddi
in Melbourne with a message telling Sussy in Fremantle to tell me to
call Anni in Claremont. Complicated? To Digital Immigrants like
you and me, sure. To them? Don't make me ROFL. The total trans-
hemispheric operation took less than three minutes.

Our collective fear of being out of touch, once I'd euthanized the
iPhone, was in this manner soothed. Sure, the kids were frustrated
initially—"Where *were* you?" the girls demanded to know when I
reeled in at 10:45 after a raucous night at the West Australian Ballet—but
they quickly grasped the ancient wisdom that "no news is good news"

and turned their terrifyingly agile minds to more productive matters. I gave them the kind of old-fashioned reassurance people have been giving their kids down the pre-mobile millennia: "If I'm five minutes late, I'm just stuck in traffic. If I don't pick up the landline at work, leave a message on my voice mail." Why this needed to be made explicit is a great mystery, but it did.

There was only one true emergency during the Winter of Our Disconnect. I got the call from Bill on my office phone, and it began with the words every mother dreads: "Mum, I've had an accident."

He'd been riding his bike, fast, down a rainy street when the front wheel detached, pitching him headlong over the handlebars and into the street. He was pretty sure he'd sprained his right wrist—a week before leaving on a European tour with his water-polo team. "Stay right where you are," I ordered. "I'm coming this second!"

"Honestly, don't bother," he replied. "I'm home now." In his words: "Some lady picked me up and took me to her place and put ice on my arm. Then she put my bike in her trunk and drove me home."

"Some lady?" I gasped. "*What* lady?"

"Dunno. Forgot to ask what her name was."

When the twitching subsided, I reflected that Bill's decision not to ring me—on *any* kind of phone—until he was home, cleaned up, calmed down, and safe was a good call. I couldn't have gotten there in time to do anything meaningful anyhow. (Thank heavens there was somebody else who did, even if we *will* never know her name. Grrrrr.) As parents, we almost think of our phones as charms or amulets: as if just holding one in our handbag is sufficient to ward off danger or repel evil forces.

What makes it tricky is that the My Cell Phone/Myself Fallacy contains a substantial grain of truth. Parenting in the age of mobile communication really does allow us to be there for our kids in a way never before possible. I don't think any of us needs reminding of that. This enhanced closeness is a good thing, largely. But like most technological gifts, it comes with big strings attached. Or maybe scotch tape is a more accurate image, being so much harder to see with the naked eye.

You know that announcement they make at the start of plays and concert performances, asking you to please switch off your phone for the duration of the event? Sometimes I think we should see our children's young years as a live performance too. As parents, it's hard enough to tread that fine line between concentration and codependency, without the added distraction of dueling ringtones. Phones don't make the world a safer place for our kids. But they are capable of fostering the illusion that we can be there—always—to ensure their safety. They are also capable of fostering the illusion that it is appropriate to try. The truth is, to a very significant extent, we can't, and it isn't.

At Seton Hall University in New Jersey, the head of family orientation—a title that in my day would have made as much sense as "alcohol-free frat party"—cautions parents against unintentionally encouraging their kids to become "dumping monsters." "The student calls home and dumps all their problems on their parents," she explains. "They say, 'This place is awful and terrible and I can't stand the food' . . . and then Mom and Dad stay up all night worrying and the kid goes out to party."[17]

At the University of Vermont, a program called "Parenting from a Distance" was recently trialed as a way to help parents through the stages of separation anxiety—their own!—when their children start college. UVM also hires student "bouncers" to keep parents away from events that don't concern them. Orientation and registration, for example. College counselors say it can be tough getting parents to butt out of roommate issues. In fact, all kinds of boundary issues are challenging now that new students arrive packing a bare minimum of five to six noise-emitting devices.

"Parental obsession with contact masks empty communication," noted one observer bluntly. "That's not love," he concluded. "It's management."[18] (The same writer noted a trickle-down effect evident among younger children, citing a *New Yorker* cartoon of an irate-looking toddler in a stroller, barking into a tiny cell phone, "I'm in the Maclaren, where are you?") After all, as Socrates might have reminded us, the unexamined iPhone is not worth having.

In the world of social networking, the term "oversharing" describes the practice of telling too much, too often, to too many. Obviously, oversharing is a judgment call—and in a world of constant technological innovation, the boundaries are being drawn and redrawn all the time. A couple of years ago, any person over the age of twelve who updated his MySpace mood more than once a day had definitely crossed the line. Since then, Twitter has raised the bar—or perhaps lowered it—elevating to an art form the posting of life's pointless minutiae.

If you are inclined to disagree, go to Twitter.com and try running a "see what people are saying about . . ." search using the keyword "soup," as I've just done, uncovering breaking news such as, "Feeling sick! Mum is making me some soup" and "Had soup today in a tea shop in Malvern called The Kettle Sings. Soup was pepper, tomato & orange. Vg—very good. Vn—very nice." Glancing up five minutes later, no fewer than ninety-four *new* results for "soup" have been posted. And it's not even lunchtime!

The reigning queen of oversharing is arguably self-described "lifecaster" Justine Ezarik, aka iJustine. Ezarik's Twitter following as of July 2009—386,000—almost matched the population of Tulsa, Oklahoma. (Admittedly, that's not even within spitting distance of top-twit Britney's 5.1 million, approximately the population of metropolitan Atlanta.)

The twenty-five-year-old also has her own TV channel: ijustine.tv. Until recently, Justine Ezarik was streaming her life live, 24/7, with a video camera strapped to her head. She stopped, she told *USA*

Today's Maria Puente, only because it was upsetting her friends and colleagues. I suspect you will not be surprised to learn that Ezarik also has a few issues around her cell phone. "I have to make a conscious effort to keep my phone in my purse and check it less," she told Puente, "but it's hard to do in the digital world we live in."[19]

Along with the entire population of Tulsa, I hear that.

Psychologist Hilarie Cash lost her son to a technology dependency. Today she heads up Internet/Computer Addiction Services, a clinic that runs treatment programs for people at risk of developing serious media-related dysfunction. "All addictions have certain patterns," explains Cash. Most of these are behavioral. Some of them are chemical, such as the release of dopamine and other opiates in the brain. The warning signs of technology addiction, perhaps unsurprisingly, are very similar to those of drug addiction: experiencing a heightened sense of euphoria while "using," and suffering extreme cravings when deprived; neglecting friends and family, and being dishonest about one's habit; withdrawing from other, once-pleasurable activities; undergoing changes in sleep patterns; being haunted by feelings of guilt, shame, anxiety, and depression.[20] Getting naked with your BlackBerry. Okay, sorry.

Experts say a person gets hooked on something like a smartphone through the process of intermittent reinforcement: that little ping of satisfaction you feel when an e-mail or an SMS arrives with the good-news message that you do, in fact, exist. That you are, in fact, still in the loop.

The specific behaviors associated with incipient smartphone addiction include checking your device when you know you have zero messages—and justifying it by telling yourself you're simply doing your job. Texting and talking while driving—a practice that *Chicago Tribune* columnist Mary Schmich suggests "is the new drunk driving."[21] Sending and receiving e-mail when walking through a crosswalk. (Thank God for "E-mail 'n Walk," a 99-cent iPhone app that uses the device's camera to act as a third eye: showing you everything you're missing, live, while you've got your nose in your e-mail.)

I did these things—and more. Yet the only true test of an addiction is the answer to the question: How well do you function without a fix?

To my surprise, my own answer was, "Perfectly."

Maybe I did love iNez not wisely but too well, but I clearly wasn't a junkie. As the days and weeks of my unplugging flew by, it occurred to me that my iPhone issues were about as hard-core as my morning-coffee ritual. Sure, I feel sorry for myself when forced to go without. I've even been known to whine, "I need coffee to function," but I'm aware (most of the time) that's just a figure of speech. Going without coffee now and then never makes me jittery or anxious or vague at all. In fact, after the initial stab of disappointment, I generally have a cup of tea or cocoa and forget all about it. I compare this to the way I felt when I was a smoker—when I would happily go without food or drink or rest or breath to procure a cigarette—and the difference is as clear as a liquid crystal display.

The truth is, once I dumped iNez, I hardly spared her a backward glance. Yes—after all we'd been through together! It was very weird, and very, very unexpected. It was like breaking off a long-term relationship and realizing, once you'd had your big cry, that you were actually perfectly fine. You almost feel guilty about feeling so good.

John Naish, author of *Enough: Breaking Free from the World of Excess*, gave up his cell phone when he realized he was becoming like the Old Man in Hemingway's *The Old Man and the Sea*. "You suddenly find yourself on one end of the line with a massive fish on the other. It feels like a prize, but it drags your little boat way into the middle of the ocean. Still you keep hanging on. And then it turns out that the huge fish is no use anyway."[22] I felt that way too. Once untethered, I experienced such a heady lightness of being. There really was nothing to crave.

The whole thing reminded me of how Anni gave up her pacifier collection at the age of eighteen months. At that point in her life—though she could speak in long sentences, recognize most colors, and recite *Goodnight Moon* backward—Anni couldn't sleep without handfuls of pacifiers being thrown like confetti into her crib at bed-time. Even then, she'd often wake me at three a.m. with a specific color request. ("Pink one!" she'd shriek, or "Glow dark!") When her favorite stuffed animal also started developing preferences, I decided it was time to call in the big guns.

One morning I announced that the Dummy Fairy would be paying us a visit that night and that she would collect all of Anni's pacifiers—for recycling, I explained—and leave her some special toys as a way of saying thank you.

The next morning, the Dummy Fairy was as good as her word. And that—to my stunned and grateful amazement—was that. Anni picked up her new doll and stayed "clean" ever after. The scourge that had destroyed my sleep and spirits for so many long months was vanquished overnight.

So, who knows? Maybe it's genetic.

The American Journal of Psychiatry added "Internet addiction" to its list of mental disorders in March 2008, reflecting interna-tional evidence of "excessive use of the Net followed by anguished withdrawal."[23] But other observers object that addiction is at best a metaphor . . . and not a very useful one at that. In his 2009 book *Cyburbia*, British critic James Harkin argues that people who are cheerfully welded to their BlackBerries, iPhones, and Sidekicks are not really addicts, but "a new kind of self"—a self that defines itself primarily through the act of navigating endless loops of information. It's a view that takes its inspiration from cybernetics, the "science of

communication and control" first articulated by twentieth-century
mathematician Norbert Wiener. Wiener was among the first to con-
ceive information (and the devices we use to deploy it) as constituting
a kind of ecology. Not a substance, addictive or otherwise. An environment. So the danger is not as much getting "hooked" as getting
lost. "Cyburbia" is the name Harkin gives to that state of limbo. Reading his book, I recognized its landmarks immediately. After all, I used
to live there myself.

It's hard to resist a book that opens with the line, "The first time I
began to wonder about our whole approach to understanding digital
communications, I was having sex on Second Life." (I have an even
more shameful confession to make. I've watched my stepdaughter,
Naomi, having sex on Second Life. But only after I begged her.) But
what resonated most for me was Harkin's discussion of "the loop"—
and how being in it can put a stranglehold on our lives.[24]

Back in 2004, Internet giant Yahoo! cosponsored a study in which
twenty-eight individuals in thirteen households agreed to go without
Internet access for two weeks and keep diaries of their experience.
Which is a bit like Thoreau going to the woods for a long weekend,
but whatever. The published extracts make fascinating reading anyway. "Every day without the Internet is frustrating," noted one participant. "I miss the private space the Internet creates for me at work,"
pined another. One guy even whined that he felt "inconvenienced
at having to carry around the paper, which was very cumbersome."
Another swore, "I'm even looking forward to seeing spam." But the
most oft-repeated refrain from the so-called Internet Deprivation
Study was the distress people felt at being "out of the loop." And those
were the exact words they used. "I feel out of the loop."[25]

I knew that exact feeling. In a very primal, playground sort of way,
my iPhone had me feel as though I was part of an in-group. Most
obviously, I guess, it put me in the loop of iPhone users: tech-savvy,

design-conscious "early adopters." The owner of a smartphone, as opposed to a . . . well, a dumb phone, I basked in its reflected glow as if it were a gifted child, and I the mother clever enough to have single-handedly engineered it.

Identifying on a personal level with the hunk of plastic you use to make phone calls is completely idiotic—and horribly, horribly human. We are meaning-makers by nature, even in situations where there is honestly no meaning to be made. Which is why we "read" our mobile media like tea leaves, or Rorschach ink blots. Secretly or otherwise, we see our devices as extensions of ourselves, just as we see our cars and our coffee tables and our kids. And we judge others that way too, in small but embarrassingly insignificant ways.

"You can tell a lot about a person by the kind of phone they carry," advises image consultant Doris Klietmann.[26] Well, duh, Doris. We personalize our phones by giving them names, for crying out loud. We dress them in leather and buy them jewelry. We think hard about their other accessories, such as ringtones and desktop photos. (My own favorite ringtone was a voice recording of my kids, then in primary school, screaming "Mummy! Mummy! Let us out!") Even the people who refuse to play the game are playing the game. My friend John, for example, is still carting around a nineties cell phone the size of a meatloaf. John says he doesn't care—that he's immune to phone shame—and I believe him. "That's just the kind of person I am," he explains with a shrug. Which happens to be exactly the point.

Being stripped of my iPhone didn't exactly precipitate an identity crisis, but I did find myself telling other users, a little too strenuously, "I used to have an iPhone too!"—just so there was no mistaking my membership credentials. Despite my temporary, self-imposed exile, I wanted the world to know, I was actually still in the loop. Really, I was.

Sussy tells me the technical term for this is "try-hard."

My expulsion from iParadise meant I had to navigate outside the loop in more substantive ways, too, reinforcing—or perhaps simply revealing—how "out there" my global positioning actually was. I'd relied on my iPhone as a kind of feeding tube, delivering a steady stream of updates from the world I was craving to connect with. The podcasts, e-mails, and applications from Up Over that I devoured helped to keep me satisfied. But they also nourished the delusion that place—actual, physical place—could be rendered irrelevant. Could be repealed by information. And, even worse, that that was okay. That it didn't matter much where you really lived, as long as you were at home in your head. Between the earbuds of your mind.

Without the iPhone to fill the empty spaces, I felt uncorked. Not depressed. More sort of decanted. Drained. At the same time, as much as I'd loved the sensation of carrying the world around in my pocket, I'd forgotten how heavy it could get. The obligations that being in the loop exacts—to be perpetually on call and responsive, to keep *up*, for God's sake—are oppressive too. And, rather paradoxically, exclusionary. The messages that aren't getting through while you're checking e-mails in the bathroom, or taking calls from under the beach umbrella, or watching a video clip of *Anna Karenina* on the station platform are something you never think about when you're inside the loop.

Sometime around month three, Sussy had declared with characteristic intensity that it was "impossible" for her to walk to the deli for a carton of milk without her iPod. "Then ride your bike," I suggested icily. But the truth was, I understood how she felt. In the first couple of days, walking pod-less for any distance greater than from the front door to the driveway weirded me out too. With nothing to listen to, the prospect of taking my usual walk to the beach with Rupert—something that normally wagged both our tails—loomed as more of an endurance test than a treat.

Just as my grandmother used to say she felt naked without her clip-on earrings, initially I felt naked walking down the street without my headset. But my ears adjusted swiftly to the shock. They started accepting other inputs: wind, traffic sounds, birds, the clatter of Rupert's tiny claws on the footpath, even—disorientingly at first—the sound of my own hair rustling. The ocean itself, audible as roaring white noise from a surprising distance away. Other sensory channels were also opening up and starting to receive signal. Those almond-shaped optical fibers on my face, for starters. It turned out there was actually quite a bit to see once you opened them up for business.

Taking exactly the same route I'd traveled hundreds, possibly thousands of times over the years, I started seeing details I'd never noticed before. A vegetable garden planted like a well-kept secret atop a limestone outcrop. The sign outside a big old house that read "T. Dire Lodging House" (a boardinghouse?! In my neighborhood? In my *century?*). Some adventurous child's teddy bear peeking down through the branches of a tall tree. A house, featureless on first glance—its brick façade the color of cheap foundation—sporting a weather vane of a matador and a charging bull. A derelict driveway haunted by the rusted-out ghosts of Jaguars past. A water feature in which a dozen long-legged, van Gogh–worthy irises stood wavering, as awkward as debutantes. A Jack Russell resting his elbows on a soccer ball. Like, who knew Jack Russells even *had* elbows?

I'm not saying my powers of observation magically went into turbo-boost. I didn't suddenly develop some poetical acuity that enabled me to see eternity in a grain of sand. I didn't spend hours eyeing squirrels and loons, or doing play-by-play for ant wars, as Thoreau did at Walden Pond. ("They fought with more pertinacity than bulldogs. Neither manifested the least disposition to retreat. It was evident that their battle-cry was 'Conquer or die.'") I had my kids for that.

But placing myself deliberately outside the loop was an eye-opener, nonetheless. In every sense.

April 1, 2009

Allowance day via automated transfer. S. blew the lot on new phone: ninety bucks got her "the second cheapest Nokia" (as first cheapest out of stock) and ten dollars' worth of credit (200 texts or 200 brain cells, whichever comes first).

First SMS to Andy in UK, who replied, "Please, please tell me this is Sussy!" Sweet! Still struggling to get my black-and-white head around concept of trans-hemispheric texting. "I don't get it. How can you afford that?" I demanded to know. S. glanced at Anni and smiled an indulgent, what-do-you-expect-on-the-dementia-ward? sort of smile. These Rip Van Winkle moments are killing me.

April 2

S.'s phone credit GONE. Finito. 200 texts in 24 hours.

April 5

Living-room coffee table, once so "adult" (expensive mags, coffee-table books, artful objets . . .), now a cross-generational mish-mash of Scrabble, sax music, pads and pens, textbooks, empty Coke cans. I like it.

Funny how nobody ever "lived" in the living room before. Then again, we never "familied" in the family room either. We surfed there.

S. and A. heading out now for McDonald's to do homework. WTF?! I explode. Apparently they have Wi-Fi there, and anyway—as everyone keeps reminding me—"THIS WHOLE THING WAS YOUR IDEA!" It's a big event now, homework. Fake tan. Outfits. Extensions.

Beg to be driven there. ("We ARE doing homework, Mum. And after all . . . THIS WHOLE THING WAS YOUR IDEA.")

April 8

New sax endurance record: four hours (between lesson, practice, and jamming with friends). Improvising to "Cantaloupe Island" tonight, B. morphed from kid with talent to pro. It wasn't practice. It was . . . music.

Parent-teacher conference at S.'s school. Hate these. The place seethes with f/t mums with manicures and tennis-club calves—earnest, high-earning couples with luxuriant heads of Liberal Party hair and high-thread-count daughters. Literally feel like a migrant.

Met S. in library, where I discovered her writing her English essay. OMG. (Usual work mode: propped on pillows in bed, surrounded by drinks and snacks, her hands flying over the keyboard like some invalid virtuoso.) All teachers spoke of "big improvement" in work habits and results since the start of term—i.e., two weeks before being rendered screenless. Coincidence? I think not.

April 9

Last day of term. S. slept from two (when she got home) to six, when I did. Then out for junk food—an event to mark the start of holidays—and home for Scrabble (S.'s suggestion).

"Is it okay to use this ancient dictionary?" S. asks dubiously, picking up our big red Webster's. Did she think it would be in Elvish or something? "It's hardly ancient," I point out. "It's a paperback. I got it in high school."

Long pause. Could see them looking at each other as if to say, "Exactly . . . loser."

Lasted three turns each. B.'s attention span particularly worrying. "Is CEM a word?" he kept asking. I reminded myself to keep
nose-breathing.

Later, S. hauled out old bag of photos for an impromptu session
of Facebook 1.0. They were mostly b.c.—first marriage, first home in
Australia, early couple, friends, etc. Stunned and almost bewildered
by the evidence that I HAVE LIVED MY WHOLE LIFE HERE.
That I was A BABY when I arrived. A tall baby, admittedly. I thought
I was so old back then. I thought I knew everything. I'd never even
heard of my pelvic floor.

April 10

Good Friday, or "Crap Friday," as Bill has suggested rebadging it.
HUGE holiday here in the world's most secular nation, for reasons
lost, like the holy grail, to the mists of time. It's the kind of day, alas,
that media were made for.

Figuring out what to do with literally nothing to watch OR buy
a challenge even for me. Eventually decided to redefine myself as
"sick" (as opposed to simply "unpleasant") and took to bed to read
and drowse.

B.'s ill temper approaching Old Testament levels—so young and
yet so grumpy!—relieved when he lit off early for Matt's after a fight
with me in which he demanded $50 worth of reeds. Returned early
afternoon with renewed restlessness.

"I need technology," he moaned. I nodded sympathetically and
patted his back.

"Do you want a hot water bottle?" I asked. The look he gave me
was not pretty.

Eventually resorted to reading to him, as he claims to be "between books." *Lolita* evidently defeated him after a few chapters. "The French bits were annoying. And I like books where something happens." LOL!!

Eventually handed over my beloved Sedaris. Anything for peace!

April 11

Did Easter shopping, dyed eggs, and made chocolate bunnies with Suss. S. and A., in their boredom, sang karaoke hymns and washed each other's feet—"What? We learned it at school"—and convinced Bill they'd cowritten the Sydney Carter hymn "Lord of the Dance," which he promptly pronounced "crap."

April 12

Easter Day. Welcome happy morning!

"Only that day dawns to which you are awake," etc. Suss up at eight-ish, came in bed with me to read papers (she reads papers now, speaking of justification by faith) and leaf through *God Stories*—inspiring tales of divine intervention, and a ripper read. At nine a.m. she reminded me it was time for church.

She is messing with my head now.

Walked to St. Paul's—gorgeous, hot morning, mercifully brief sermon—then back to discover A. & B. happily munching basket contents for breakfast, B. having written me a "sorry" note for yesterday's sullenness, complete with tacky Jesus sticker. Later, in honor of the day's festivities, allowed girls to take videos of themselves dancing. All sat down and attempted to play Tiddlywinks ("It's hard!" Suss cried).

After lunch, when Mary and kids came, Sussy and Torrie (aged fifteen) got out the watercolors, while Anni and Ches walked to the beach with B., who called later to request a sleepover (and enjoy brief resurrection experience with the DS, no doubt). Filled with the spirit, and quite a bit of Semillon Sauvignon Blanc, I said yes.

April 14

From *The Telegraph*:

"The fifty-year-old singer, a known health fanatic, would spend up to four hours a day in the gym before taking her children to Kabbalah religious meetings, according to friends. Breakfast was low-fat smoothies and suppers would consist of steamed fish and vegetables, while television was regulated because Madonna believed it harmed her children's development."

S. excited to learn that Madonna's kids don't watch TV either. "I didn't know you liked Madonna," I say. "I don't," she replies. "She's a creepy loser."

Sometimes I feel exactly the same way, minus the compensatory thigh muscles.

April 16

Bill and Suss fistfighting over *Lolita* access even as I write. (NB: Suss takes French and now that she wants to read it, Bill has suddenly gone all postmodern.)

Monopoly tonight with B.'s friend Jake, B., Torrie, and self (S. painting toenails in the peanut gallery) on battle-scarred board: Only original token left is sports car (stand-ins include old Parcheesi piece, wishing stone, button, and Trivial Pursuit pie slices)—at least I don't get stuck with the iron anymore—and half the deeds are missing too (some replaced with handwritten facsimiles circa 1998—others simply memorized).

They say Monopoly reveals who a person truly is. Oh, God. Not that.

Bill morphs into consummate capitalist hyena, complete with Foghorn Leghorn accent ("Two hun'red dollahs? Ah say, ah say, 'Kitty litter' son.") Only Anni can Trump him at his own game—and I do mean Donald.

Am not unsympathetic. I too went through a Monopoly phase myself back in the summer of '68 (till my folks called in the National Guard). Pretty much ever since, though, have nursed a secret loathing of all board games. So the truth is, The Experiment hasn't so much forced the kids to play Monopoly. It's forced ME to play (i.e., "THIS WHOLE THING WAS MY IDEA").

Shocked that we had a blast. Especially loved it when Bill called Torrie at 8:15 p.m. to ask if she wanted to come over and play, and she arrived on our doorstep at 8:16. Afterward, B. and Jake (age sixteen) played Connect Four and built a tower of wooden blocks. What? No wading pool?!

April 20

Friends freaking re: my iPhonelessness. Twice this week have met anxious-looking loved ones for coffee. "Wasn't sure you'd remember, and I had no way to contact you!" Interesting. Since when did it get so hard to remember a coffee date without appliances? "I *am* allowed to write stuff down," I reminded them a little huffily. To be fair, I was ten minutes late on both occasions. But then, I'm *always* ten minutes late. A text to say so would be redundant.

Suss on landline to Maddi for 2.5 solid hours.

A. reports The Experiment a popular discussion topic at Sunday pub session. One guy her age admitted he'd spent ten consecutive hours playing an online game that day. Says you know you've hit rock bottom when you can see the Tetris blocks falling on the inside of your eyes when you shut them to sleep.

Pointed out (as if I hadn't noticed) that her screen-free room is much neater now—less a giant laundry hamper, "more of a haven" is how she put it.

B. home early from Fairbridge Music Festival, where has been camping with friends. Evidently forgot to pack clothes, pillows, shoes, or food. Oops. Asked humbly for clean sheets and sank onto them, book in hand, cool jazz on the stereo.

April 24

Bill played Satie's *Gymnopédie No.1*, on the piano, beautifully.

Later, he came home from Pat's house with sheet music. Had spent the day learning it. I was floored. Practiced sax scale drills (major-minor-harmonic-melodic) in kitchen while I cooked.

Overwhelmed with admiration, wonder . . . and guilt. If we'd been "experimenting" all along, where would he be now??

April 26

B. got a job today!! And I thought the Satie thing was a miracle. Fif-teen dollars each and every afternoon after school to vacuum and mop a very small café. He is rich. I am happy.

A. and self on couches tonight reading (*Why Men Marry Bitches* and *The Book Thief*, respectively) and doing *New York Times* cross-words from 1987 (unearthed from an ancient diary).

April 27

S. to Angel Café—suitably gowned and coiffed—to work on personal project (is creating a magazine for teens . . .). There for FOUR HOURS! When asked how much of that actually spent on homework replied, "Probably half."

Lil sleeping over. She and S. in family room playing Monopoly, eating Tim Tams and drinking cocoa, and spritzing kitten with a spray bottle to keep her from pouncing on hotels.

S. tells me all her friends are watching the not-yet-released Hannah Montana movie on Sidereel but she has taken a vow of chastity (as it were) and will "save herself" till the cinema launch.

On football permission form, B. filled out his name as:

Bill.Christensen

and mine as:

Susan.Maushart

What's up with the dots? I ask him. What do you mean? he replies, puzzled. Isn't that the proper way?

In other punctuation news . . . Sussy's friend Sean—completely OCT (obsessive compulsive texter)—now says the words "question mark" at the end of spoken questions. "Are you coming to the party, question mark?" he'll ask, complete with rising inflection.

The Sound of One Hand Doing Homework

The greatest menace to progress is not ignorance,
but the illusion of knowledge.
　—DANIEL BOORSTIN, *Cleopatra's Nose*

On a distant galaxy (like, six months ago), in the space formerly known as the family room, three teenagers are doing their homework—a task that requires wireless broadband with unlimited download allowance, six GBs of RAM, a terabyte of hard-drive space, five cell phones (two are just spending the night), three iPods, two printer/ scanners, and a color cartridge in a pear tree.

Everyone knows the Internet is a powerful research tool. Maybe that's why using it to do homework is like exfoliating with an orbital sander. Like, sometimes it's not good to go so deep. When all you want to know is why the Boxer Rebellion failed (and no, darling, it was not "something about the elastic"), and you get 32,700 results . . . frankly, the mind Googles.

Sometimes it seems providing Internet access so your kids can do their homework is like using a vibrator to whip cream. It's not only inefficient. It betrays a serious want of imagination.

Consider Sussy, ostensibly toiling away in there at an essay on

e. e. cummings. At the moment, she has no fewer than nine windows active on her laptop. Six of them are online conversations that, leaving aside the missing punctuation, uncertain syntax, and sophomoric self-absorption, are off the topic entirely. A seventh is illegally siphoning off the latest episode of *The Secret Life of an American Teenager*, while an eighth is tracking an online auction for a pair of . . . weirdos? "Weiros, Mum," she corrects haughtily. "They're birds."

Oh, birds. Well, that makes sense, then. On the table, her cell phone, set to silent, squirms helplessly. I watch it with something akin to empathy.

Sussy—along with pretty much every other kid alive—insists that multitasking rules. The research sends a different message entirely. Although the latest neuroscientific reports confirm that media are changing the very structure of our brains, substantive evolution, like homework itself, is a painfully slow process. If Sussy were still writing that essay on e. e. cummings in 500,000 years (and at the time, the odds looked promising), she might have a good case. But for now? In our lifetimes? There isn't the slightest doubt that monotasking kicks it.

Your kids will tell you otherwise. They're not necessarily lying either. They may be absolutely convinced they can work just as well with half a dozen conversations and a couple of cyber-weiros squawking in the background as they can with silence. But these are the same people who think Micronesia is a software developer. Do we really want to take their word for it?

The Greek root of the word "technology," *techne*, means "skill," or "trickery." Yeah, well. Sometimes it *is* hard to tell the difference. Maybe that explains why I was so often in a state of high alert watching the kids doing their, quote-unquote, homework within the media bubble our family room had become. Not that my anxiety ever did any good.

Fretting over what, exactly, they were studying so intently—or with whom they were studying it—was my problem. Or so Anni would remind me, on those rare occasions when she'd break suction, visually speaking, from her laptop.

She was a straight-A student, give or take. Okay, so maybe there'd been a bit more take in the last couple of years. Year 12 had been spent almost entirely on exchange. At the time, that's how I preferred to think about MySpace. Her absenteeism took its toll on her final grades, but she'd still done easily well enough to get into the university of her choice. Quibbling that the ninety wasn't a ninety-seven was ungracious, but I did it nonetheless. "It's not the grade that bothers me," I'd stammered at the time. "It's the . . . the effort."

"What effort?" she wanted to know.

Well, exactly.

"But imagine the results you could have gotten if you'd tried," I pressed. "If you took all the hours you'd spent on MySpace and MSN and poured them into Lit, or French." She shrugged.

"I would have gotten higher test scores, I guess. But I would still have gotten into the same university, and I would still be majoring in journalism."

"Can't you think of any other differences?"

Finally, I could see a light dawning in her eyes. "Oh, I see what you mean," she said. "Of course. Why didn't I think of this myself?"

"I wouldn't have had a life."

I reminded myself at such moments of William Morris, the nineteenth-century English designer and visionary. Morris liked to stage personal and completely ineffectual protests against technological excess too. Once he sat on his top hat to show contempt for the Stock Exchange. I'm sure it made him feel better, briefly, too.

Morris was no Luddite. In fact, he made a fortune mass-producing his designs for fabric and wallpaper. But, also like me, and probably

you too, he was convinced that "if we hand over the whole responsibility of the details of our daily lives to machines and their drivers,"
happiness would elude us. People needed to learn how to restrain
themselves around technology, he believed—in fact, around all household stuff. "Have nothing in your house that you do not know to be
useful, or believe to be beautiful," he famously admonished.

At one level, most of the gadgets my children were employing to do
their homework fitted Morris's criteria to a tee. The minimalist Mac-
Book with its sleek lines and whisper-soft keyboard; the twinkling,
tinkling little Nokias; the petal-pink Nintendo DS with a stylus as
delicate as daisy stem. Considered singly, they really were useful and
beautiful. It was when you threw them all together and stirred them
around that things started to get ugly and pointless. Not to mention
chaotic and cacophonous.

It was so very foreign to the way I myself work best: in monastic
seclusion and pin-droppingly pure silence. Even the most innocu-
ous background melody grates like fingernails on the blackboard
of my attention span. Music with lyrics can make me scream with
frustration. When we had an extension built some years ago, I grew
accustomed to the racket of the power tools. But the talkback on the
tradesmen's radios nearly killed me. In the end, I was driven to using
earphones on my own radio, tuned to static. As for television, having
it on in the background while I try to write—or even read—is almost
physically painful. At the first few bars of the *Simpsons* theme song, I
swear I can hear my ideas snapping like twigs.

"Can't you just block it out and pretend it's not there?" plead
the kids.

"Absolutely," I say, advancing on the "Off" button. If I can't hear
what I'm writing, I try to explain, it's no longer creating prose. It's
tossing word salad.

My children naturally regard my work habits as freakish. But they are fairly common among Boomers. In our day, people still sat down to watch television. It was an intentional thing, and it usually involved a program guide. There were no remotes. There were no DVDs in people's bedrooms. (That, I explain to my wondering children, would have been like having a gas barbecue next to the bed, or a flush toilet in the closet. "Sick!" they chorus.) Television had not yet become the soundtrack for family life—the not-beautiful, not-useful wallpaper lining every household's personal space; the visual cud on which the entire culture chewed.

"But Mum . . . We *need* the background noise," they explain. They speak slowly, as if *I* were the one with the intellectual impairment. As, arguably, I may be.

Sussy confides that one of the reasons she loves her school is the teachers allow iPods during class. "During CLASS?!" I sputter. (For this educational privilege I'm refinancing the family home?)

"Well, yeah. Why not? When we're doing projects and stuff."

I tell her about the time my father yelled at me for humming under my breath at the dinner table. I tell her about my primary school cafeteria, where you weren't allowed to talk. I tell her about the fitness revolution sparked by the appearance of the Walkman in the late seventies. ("The idea that you could go jogging and listen to music at the same time blew people's minds," I explain. "Random!" she cries with real feeling. "But I don't get it, Mum. What's 'jogging'?")

No wonder my generation doesn't do well with a lot going on in the background—not cognitively, at any rate. For us, as for Gerald Ford, who famously struggled to chew gum and lead the free world, doing more than one thing at a time is a reach.

The multitasking millennials we have spawned may be fully functional with a device in every orifice, or they may not. But we, their Boomer and Gen X parents, know our limits. We excel at monotasking.

Our kids smile indulgently at our limitations. It's as if our inability to multitask were some quaint but faintly repulsive evolutionary vestige—like hairy knuckles or a monobrow. When we thunder, "But how can you THINK with that racket going on?" they explain sweetly, "The thing is: Our brains are different." (I don't know about yours, but my own kids seem to have learned neuroscience before their times tables.) They don't say "superior," or "more highly evolved," but that's what they mean, and they and I both know it.

Bill's idea of dressing for dinner is wearing board shorts that Velcro all the way up. Shoes? That's pushing it. Anni, who prides herself on her immaculate grooming, sheds her hair extensions as unthinkingly as a snake sheds its skin. Yesterday I noticed them on top of the piano, snarled like roadkill. Sussy pours cereal directly into the drawer, where, most of the time, it hits a bowl.

Believe me when I tell you: Their brains *are* different.

However, I'm pretty sure that's not what they're getting at when they assure me that multitasking is a piece of cake (when, to me, it looks like a mangled collection of crumbs). They probably mean more or less what Don Tapscott says in his buoyantly optimistic book *Grown Up Digital*. "Parents have said to me: 'How can my kid be doing homework, while he's also listening to MP3 files, he's texting on his phone, he's got three windows open on the computer—one of them Facebook—and he's petting the dog? How is this possible?'" Tapscott's answer is scarily identical to Bill's. "Well," he says, "it turns out they have different brains."[1]

A Canadian business consultant with an undergraduate degree in psychology and no children, Tapscott assures parents that worrying about kids' multitasking is like worrying that your mate no longer carries a club. "These kids grow up interacting and collaborating, thinking and organizing, scrutinizing, having to remember things, managing information. And that affects the actual wiring, synaptic connections, and structure of kids' brains. So they have better switching abilities and better working active memory. If I'm doing several things at once, I can't keep track of what the heck is going on, but they can. So this is creating a generation that thinks, works, and learns very differently."[2]

The last sentence I have no problem with. About the rest, I'm not so sure. Better switching abilities? Hmm. On the remote, perhaps. Better working memory? Well, maybe—as long as we're not counting the number of cell phones, iPods, and chargers missing in action on the bus last year. Call me unevolved, but I get nervous when non-scientists start talking about the "actual wiring" in our heads.

Before we began The Experiment, my own (admittedly equally unscientific) hunch was at odds with Bill's and Tapscott's. The idea that constant track-jumping could conduce to a smooth train of thought for anybody, under any circumstances, seemed like a stretch.

Post-Experiment, I know it is. I watched as my kids awoke slowly from the state of cognitus interruptus that had characterized many of their waking hours, to become more focused, logical thinkers. I watched as their attention spans sputtered and took off, allowing them to read for hours—not minutes—at a stretch; to practice their instruments intensively; to hold longer and more complex conversations with adults and among themselves; to improve their capacity to think beyond the present moment, even if that only translated into remembering to wash out tights for tomorrow morning.

I'm not saying anybody suddenly went from Hannah Montana to Homer. They didn't develop an unquenchable thirst for their set texts, or learn to love their trigonometry worksheets. In fact, they probably did no more homework during The Experiment than they had done before—Sussy swears she did much less and, though her grades improved significantly, this may be true. But they all completed their schoolwork far more efficiently, far more quickly, and with visibly greater focus.

I can't say what went on in the "actual wiring" inside their heads once they were forcibly encouraged to monotask. Luckily, I don't have to. Any number of major studies has shown that all that spin about "different brains" has been greatly exaggerated, like Mark Twain's death or the redemptive power of bleached teeth. In fact, it turns out that multitaskers are not ahead of the cognitive curve at all—not even in those skill areas where one would expect otherwise. It's true that our kids' brains are being changed by the media they habitually interact with, and that many of those changes are as yet dimly understood. It's also true that bookish people like me, who need cocoonlike isolation in order to work effectively, have our own wiring issues. But the fact is no one's brain is different enough to make constant interruptions, distractions, and task-switching an optimal environment in which to function. *No one's.*

As a fifty-two-year-old, post-reproductive female, my brain is "different" too. Well, hello. Half the time I can't even remember where I left my last partner, let alone my reading glasses. As we all know, the prevailing cultural mythology for people my age—especially women my age—is all about memory loss, vagueness, and diminution of brain function. At least, I'm pretty sure it is . . . Hang on. Isn't it?

LOLishly enough, the latest neuroscience suggests that people at my stage of the game have particularly agile neural ability. (Think Barack Obama, not *Menopause: The Musical*.) True, we are somewhat slower at acquiring new information. But our ability to process, organize, and contextualize that information is unparalleled—and it shows in our "wiring"—aka our neural structures. Midlife brains are marked by a proliferation of glial cells (that's Greek for "glue") and experience optimal convergence between right and left hemispheres. The cumulative effect, notes one neuroscientist, is that "our brains graduate from a dial-up modem pace to high-speed DSL." No wonder we occasionally exceed our personal download allowance!

The superior cognitive function we experience at midlife is one reason people tend not to vote for world leaders who are in their twenties and thirties. It also explains why younger air traffic controllers are consistently outperformed by their more experienced elders. A recent study by researchers at MIT and the University of Illinois found that middle-aged workers' reaction time, memory, and attentional ability was significantly worse than that of younger colleagues, when both groups were tested in isolation and under laboratory conditions. But when they were tested in real-life conditions, the elderly tortoises absolutely hammered the upstart hares.[3]

You and I might call the midlife advantage "wisdom." Neuroscientists locate it within actual structures in the brain. It all adds up to the same thing: When it comes to sorting through and weighing up multiple bits of information, midlife heads do it better and faster.

While we're on the subject of good news for mother, midlife brains are also more efficient when it comes to controlling temperament. Contrary to prevailing stereotypes, males and females alike become *less* grumpy with age. We also tend to grow less impulsive, less labile in our moods, and less prone to extreme emotional responses. A study carried out at the University of California, Berkeley, assessed 123 women in their early twenties and again in subsequent decades. It found "likable personality traits"—such as the ability to remain objective, to tolerate ambiguity, to handle interpersonal relationships successfully—peaked for women in their fifties and sixties.[4]

Admittedly, on the day of the e. e. cummings assignment, you could be forgiven for thinking otherwise. It was one of those days where something definitely clicked in my head. It was somewhere between an *aha!* moment and a *WTF?!* moment.

That night, after I'd kissed everybody goodnight and switched their screens to sleep mode, I found myself thinking back to my own high school days. I'd usually do my homework in my bedroom, away from the depressingly familiar dialogue on my mother's afternoon soap opera. Other kids tore handwritten pages from their spiral-bound notebooks, but not me. Those untidy curly edges always set my teeth on edge. I preferred to do my assignments double-spaced neatly on onionskin, on my beloved orange Olympia portable. (Once a nerd, always a nerd.) Yet there was nothing particularly Dickensian about my bedroom. I had a telephone. It was white and had a bleached-blond troll doll glued to the receiver.

I also had a radio and a stereo, and in my senior year even a portable color TV. But using any of them while I worked would have been as unthinkable as singing karaoke in the middle of a school assembly. And there *was* no karaoke back then.

I did get bored sometimes. But most of my distraction strategies seemed to revolve around fire: lighting a stick of incense or a scented candle—hey, it *was* the seventies, okay?—or, in extremis, crawling through my bedroom window onto the roof to smoke. I also had a weird but obscurely satisfying habit of melting crayons on the light bulb of my desk lamp. That was the closest I ever came to something resembling multitasking. I listened to a ton of music, like all normal teenagers. But when I listened, I listened. Hard, and usually studying the lyrics on the back cover of the LP.

Other kids I knew were pretty much the same. Some listened to the radio while they studied—something our parents and teachers frowned upon, it amuses me to remember—but that was about as "stimulating" as our media environment ever got.

These are anecdotal recollections, I realize. Yet the fact that there exist no hard data from this period on teen media use is evidence of how much has changed. Today, entire journals are devoted to the subject, and new articles and books appear as regularly as reruns of

Seinfeld. We are so much more interested in how our kids interact with technology. Partly that's because there is so much more technology. Partly it's because there's so much more fear. Thirty-five years ago, we didn't know enough to know how much we didn't know. Today we are beginning to.

You don't need a Ph.D. in social psychology to tell you something's up when you have to fight to make eye contact with your teenage children, or get them to sit down and eat a meal, or have the occasional grunt-free conversation. As parents, we comfort ourselves with the excuse that all this is normal, natural, age-appropriate stuff. But somewhere in the back of our beleaguered, Boomer-ish brains, we remember a time—perhaps even our own teen years—when it wasn't. For many of us, that is exactly the opposite of a comforting thought. It's a scary one. And it's exactly that fear factor that creates such fertile ground for the growth of noxious "experts," whether they are the cheerleaders, who insist we are moving toward a user-generated golden age waaay too cool for parents or other vestigial organisms to understand, or the doomsayers, who prophesy with equal confidence the collapse of civilization as we know it.

Among the former are authors Don Tapscott—the aforementioned "different brain" dude—and Steven Johnson, whose irresistibly titled *Everything Bad Is Good for You* argues that the new media only *seem* to be dumbing us down; in fact, they are making us smarter. True, our kids know fewer facts and less history, these authors acknowledge. They struggle to construct arguments and to maintain focus. But their ability as information hunter-gathers, their visual acuity, their narrative and creative intelligence, leave an older generation for brain-dead. This is exactly the message our kids want us to hear. It's also the one they themselves quite earnestly believe. To be fair, there is some compelling evidence to support the cheerleaders' case—including the fact that IQ (as opposed to SAT scores, say) has been rising for decades.

In the opposite corner sit observers like Emory literature professor Mark Bauerlein, author of *The Dumbest Generation* (no prizes for guessing how *he* feels about multitasking); journalist Maggie Jackson, whose meticulously researched book *Distracted* argues that, in the age of the Internet, we *all* have ADHD; and clinical psychologist Michael Osit, whose 2008 book *Generation Text* thunders against a generation used to "instant everything." The case the doomsayers argue is persuasive, fact-filled—and deeply depressing. Literacy as we know it is vanishing. Attention spans are anorexic. Narcissism is up—knowledge is down. The culture is coarsening and so is our cognitive edge.

The cheerleaders tell only the rah-rah side of the story. They fall over themselves in their eagerness to seize the new day. The doomsayers' gaze, by contrast, is fixed determinedly in the rearview mirror. They see only a rapidly retreating landscape and devote most of their considerable rhetorical power to mourning its passing.

The question of which side to believe is so pre–Web 2.0. As critic Neil Postman was fond of remarking, "Information explosions blow things up." That doesn't mean they blow *everything* up. But some stuff, yes, inevitably—sometimes some pretty important stuff. The fireworks are amazing—but we pay a price for admission. Media give, and media take away.

History shows us that, with the turning of each new technological tide, there is always somebody who'll forecast tsunami. Socrates was one of them. He feared that the written word—basically the Twitter of fourth-century Athens—would undermine education, and he warned that reading would cause people to "cease to exercise their memory and become forgetful." (Yup. The exact same argument you and I are *still* making about the use of calculators in school.) Having too many facts at one's fingertips "without proper instruction" was dangerous too, leading people to be "filled with the conceit of wisdom instead of real wisdom."[5] (The exact same argument you and I are still making about Google.)

Fifteenth-century Venetian man of letters Hieronimo Squarciaf-
ico thought the *printing press* was the devil. "Already abundance of
books makes men less studious," he fumed. "It destroys memory and
enfeebles the mind by relieving it of too much work."[6] A German
critic, writing at the dawn of the reading revolution that would sweep
Europe and the New World in the early nineteenth century, prophesied a pandemic of "colds, headaches, weakening of the eyes, heat
rashes, gout, and arthritis."[7]

In *Everything Bad Is Good for You*, Steven Johnson cranks the culture jam even louder, imagining what today's conservative critics—the
ones who are convinced that Wikipedia is the devil's workbench—
might have said about printed books. That they are "tragically isolating." That they understimulate the senses. That they suppress social
interaction and breed intellectual passivity. (Imagine! You simply
sit back and have the story dictated to you.) All true, of course . . . and
all conveniently overlooked by Digital Immigrants such as you and me.

Most of us don't think of books as "media" at all—which is both
ridiculous and a reminder of how utterly embedded in our media ecology they have become. It's sobering to realize that Socrates' version
of The Experiment would have been a six-month ban on reading or
writing. Random! It occurs to me that I should really try to do without
for a day or two, out of a sense of fair play to the Natives. The prospect
frankly terrifies me. What the hell would I *do*? How would I *keep
up*? (Wait a minute. Could this be the way Anni, Bill, and Sussy felt
about relinquishing their Facebook accounts?) The concept of having dependency issues around literacy had never occurred to me, any
more than the concept of having dependency issues around oxygen.

I'd made the appointment to speak to Bill's Year 11 adviser almost
impulsively. It was toward the end of the second term (and The
Experiment). He was doing all right—overlooking the 27 percent on

his last pre-calc exam—and he was happy enough with his teachers. But I still had concerns. I wasn't sure his subject choices were the right ones, I explained to his adviser.

"Overall, Bill's results have been excellent," she reassured me. "His two maths units, his English, his human biology, his phys ed studies . . ."

"Yes," I interrupted. "I know all that. It's just . . . I'm afraid he's going to graduate next year without . . . well, *knowing* anything."

She looked at me quizzically. "Go on."

"He hasn't taken any geography or history since Year Nine. He doesn't study a language. He doesn't read politics, or law, or litera-ture, or art history, or a social science of any kind. In English, so-called, they watch movies . . . And half the time they don't even write assignments. They make PowerPoints, as a group project . . ." I trailed off. I was remembering the time Sussy's teacher penalized her for using "too much language" in a slideshow presentation. If too much power corrupts, I'd reflected at the time, too much PowerPoint corrupts absolutely.

I continued. "He's a bright boy, and he does his work, but he doesn't *know* anything."

Bill's adviser put her pen down. "Ah," she said. "That."

In any discussion of the impact of media on thinking and learning, it's vital to distinguish between aptitude (our cognitive capacity, or "intelligence") on the one hand and cognitive style (the habitual ways we use our intelligence) on the other. That said, in recent times, both have undergone significant change. Among other things, this means that your kids really are smarter than you are—just as you suspected and they keep telling you. At the same time, they really are more impaired—just as you suspected.

Hang on. Is *this* why they can do complex file conversions in their sleep, or edit a YouTube video with one hand tied behind their Facebook account, but still can't remember where Antarctica is? ("Wait—down south, right?")

Well, it could be. Because, hard as it is to believe when watching an episode of *Project Runway*, as I've said, studies show that across-the-board intelligence is increasing, at least insofar as IQ scores can be said to measure it. A score of 100 on an IQ test is still "average." But that's only because the tests are being constantly recalibrated. Raw IQ scores show an average increase of about three points per decade since 1920. And it's not just the better fed, better educated, more affluent segment of the population who are growing sharper. Those in the middle part of the demographic curve are too—including "the people who have supposedly been suffering from a deteriorating public-school system and a steady diet of lowest-common-denominator television and mindless pop music," as journalist Malcolm Gladwell puts it.[8]

The trend is not in spite of our increasing reliance on electronic media, but because of it, argues Steven Johnson. He believes that TV, computer games, and social media all place greater cognitive demands on users than earlier forms of leisure. This doesn't seem to make sense if we compare, say, watching *Australian Idol* with reading *The Brothers Karamazov*. But how about sitting on the back step chewing tobacco? Or darning socks? Or falling asleep at sundown after a twelve-hour day on the assembly line? Most people were *never* reading fat, complex Russian novels. Most people were staring into space. And compared with that, *Today Tonight* is brain food. Or so, at least, the argument runs. At the lowest level, more time assimilating content—however puerile—means less cognitive downtime, means more neurons firing, means increased capacity. Theoretically.

If all the time Anni is now spending on Facebook were devoted instead to, say, debating the carbon emissions trading scheme or acquiring the rudiments of Katakana, the brain benefits would be clear. But if instead she were embroidering a table runner . . . ? Hard to say.

When Bill rediscovered the joy of sax, he started out playing maybe twenty minutes a day. By month four of The Experiment, he was practicing for up to three hours a day. Basically, he'd swapped Grand Theft Auto for the Charlie Parker songbook—and he knew it. "What if I could take back all those hours I'd spent on The Beast," he mused, "and used them for practice? I'd be *amazing* now." Pretending that the same thought hadn't occurred to me twelve thousand times was one of the toughest tests I've ever faced as a parent. I swallowed hard and tried to nod spontaneously. But in truth, it was still a big "if." Giving up one activity does not guarantee you will take up a more worthy substitute. The possibility that you might quit smoking and become hooked on nicotine patches instead is a real one.

Sussy's experience was a good case in point. Under the new regimen, her screen time dropped from around six hours a day to about one. (She still used her laptop at school, ostensibly for work.) But her talking-on-the-landline time ballooned to fill perhaps three-quarters of the gap. This had some interesting repercussions for her friendships (not to mention our phone bill). But the difference between IM-ing her friends on her laptop and talking to them on the landline was arguably a toss-up, cognitively speaking.

Anni was more diverse in the way she approached her newly freed-up free time: reading more, seeing friends more, cooking more. But she also spent long periods in bed, leafing idly—let's not say slothfully—through glossy magazines and listening to crappy commercial radio. Was this in any sense "better" than her pre-Experimental binges of eBay window shopping to the accompaniment of an iTunes playlist set to terminal shuffle? It's hard to see how.

Back when *I* was a kid listening to crappy commercial radio, we were taught that intelligence was a fixed entity. Like the bowl of waxed fruit on our dining room table—or, for that matter, like matter itself—intelligence could be neither created nor destroyed. Smart kids were born smart, and they would stay that way. Dumb kids sat at a special table where they belonged. Today we know the whole intelligence question is much more complicated. For one thing, we recognize that there are *kinds* of intelligence. Even our primary-school kids learn that now, and high time too. But we have also discovered that cognitive capacity can be cultivated. We know now that brains are "plastic": not in the Barbie doll sense, but in the plasticine sense of being moldable. Brain structures—neurons and neural pathways—can and do change significantly with use. Like a catcher's mitt or a nursing mother's breasts, they morph to fit the use to which they are habitually put.

To an important degree, we really do become what we behold—and so, it turns out, do our brains. It's not so much the content we absorb that makes the difference. It's the way that content is packaged and transmitted via symbols (like an alphabet or semaphore) and media (like the printing press or a Bluetooth headset). When Marshall McLuhan famously observed, "The medium is the message," that's what he was getting at. Reading *Harry Potter* may be "better" than seeing the movie, or it may be "worse" . . . but it is an entirely different story, quite literally, as far as brain function is concerned. In the same way, readers of ideograms, like the Chinese, develop a neural circuitry demonstrably different from that of alphabetic readers—and the differences are discernible across many regions of the brain, from the way memory is stored to the way visual and auditory data are processed. Ditto when you attempt to improve your tennis by playing a video game, or to cook macaroni and cheese using a tutorial on your handheld game (as Bill once attempted to do, reducing the entire household to a figurative puddle of cheese sauce).

It would be peculiar if People of the Book, as Digital Immigrants are by definition, did not develop along different cognitive lines to People of the Screen. The question is, how different—and different how? Watching my kids juggle e. e. cummings, video uploads from the latest social event ("It's not a party, Mum—it's a 'gathering'"), instant messaging conversations with forty-seven of their closest friends, and the odd spot of extreme Googling, I used to wonder all the time how they did it. Their preferred explanation—that the multitasking teen brain simply has powers and abilities far beyond those of mortal monotaskers—seemed so logical. I mean, seriously: *I* couldn't do what they do.

Imagine my surprise when it turned out that they couldn't either.

Many morbidly obese individuals eat three meals a day. It's not how often they eat that creates problems. It's what they put on their plate. Maybe we shouldn't be surprised that the epidemic John Naish calls "infobesity" works in much the same way.

The Kaiser Family Foundation's latest study of multitasking teenagers found that Digital Natives weren't necessarily spending more time with media than their parents were in the seventies—they were just packing more into it. American teenagers today spend an average 7.5 hours a day with media. But because using more than one device has become their new default setting, the figure for total daily media exposure clocks in at a horrifyingly Huxleyan 10 hours and 45 minutes. That's an increase of more than two hours in the last five years.

For Generation M, multitasking acts as a kind of cognitive wind chill factor, intensifying engagement, fragmenting attention, and transforming entirely the experience that used to be called "tuning in."[9] A *Los Angeles Times*/Bloomberg poll conducted way back in 2006 among 1,650 teens found that, while doing homework, 84 percent listened to music, 47 percent watched TV, and 21 percent did three tasks or more.[10] (And no, I'm not sure when watching Extreme Jackass stunts on YouTube got reclassified as a "task" either.) The 2010 Kaiser research found over 58 percent of seventh to twelfth graders say they multitask "most" of the time. Among eight- to eighteen-year-olds, one in three admit to multitasking "most of the time" while doing their homework.

The good news is that neuroscientific research in this area is accelerating almost as fast as your fourteen-year-old's status updates. The bad news is that findings remain pretty basic, and only a few things are known for sure. One of them is that there is actually no such thing as multitasking.

Truly. Unlike your mother, your brain really can only do one thing at a time. Or, more accurately, it can only process information one task at a time. What may look like simultaneous engagements are actually sequential ones. The mind boggles, in other words, but the brain *toggles*—sometimes quite rapidly, from one task to the next. It's an action that occurs in the region behind the forehead, the anterior prefrontal cortex, aka Brodmann area 10—a region that is one of the last to ripen (and one of the first to rot) with age. Not surprisingly, therefore, young children are actually worse at task switching than adults are.

David E. Meyer, director of the University of Michigan's Brain, Cognition, and Action Laboratory, minces no words. Multitasking, he says, "is a myth." And it always will be, thanks to the brain's inherent limitations for information processing. "It just can't be, any more than the best of all humans will ever be able to run a one-minute mile."[11] Different brains, maybe. But *that* different? Uh-uh.

Meyer's research shows that Sussy has been paying a heavy price for the privilege of Skyping her bestie while "simultaneously" studying for tomorrow's science test. Digital Natives tested in Meyer's lab took double the time or more to complete tasks while multitasking.

Even more worryingly, their errors went way up. The "mystique" (as Meyer calls it) that this generation has sought to perpetuate about itself is a hopeful, if not downright arrogant, delusion.

When you think about it, you realize that of course this would be the case. Bill was absolutely and sincerely convinced that keeping one eye glued to the action-packed anime feature playing in the background made no difference to the quality of the essay on racism he was slowly but uncertainly pecking away at in the foreground. But then he would, wouldn't he? Before The Experiment, he'd never really tried it any other way. What basis for comparison did he—do any of them—really have? The most recent research makes the point even more strongly.

A series of experiments carried out at Stanford's Communication Between Humans and Interactive Media Lab and published in the *Proceedings of the National Academy of Science* in August 2009 tested two sets of college students—those who self-identified as heavy media multitaskers and those who said they were light multitaskers—on a range of problem-solving tasks. The main finding?

"Multitaskers were just lousy at everything," is how researcher Clifford Nass summed it up. Nass and his fellow researchers set out to identify the cognitive advantages of multitasking. They'd designed the study with that purpose in mind. What they discovered was so contrary to expectations that Nass admitted, "It keeps me up late at night."[12]

Weirdly, the heavy multitaskers were especially disabled when it came to . . . well, multitasking. Compared with their peers, they were terrible at filtering out distractions. They were also less efficient at task switching, routinely paying a much higher cognitive "switch cost" in pace and accuracy.

Researchers were surprised to find that the experienced multitaskers had problems with working memory too—basically, they were less selective about what data they paid attention to, and this made them more vulnerable to distraction.

"I was sure they had some secret ability," Nass commented later. "But it turns out high multitaskers are suckers for irrelevancy."

"We kept looking for multitaskers' advantages in this study," added principal researcher Eyal Ophir. "But we kept finding only disadvantages. We thought multitaskers were very much in control of information. It turns out, they were just getting it all confused."[13]

They're not the only ones. "The core of the problem," Nass muses, is that multitaskers "think they're great at what they do; and they've convinced everybody else they're good at it too."[14]

When the kids started back to school and university in February 2009, I was more nervous about the homework question than they were. By that point, I'd had a month of trying to write in the old-fashioned way, and the results did not exactly inspire confidence. Completing my weekly column, a task that normally consumed an hour or two of online research and half a day of writing and rewriting, was now taking two full days to pull together.

Switching from research-driven topics to more reflective ones was easy. It was fun being less tied to data, using newspaper or magazine articles as a jumping-off point for my own ruminations. It was a method I'd employed intermittently throughout the ten-plus years I'd been churning out weekly copy, and I found it often simplified my writing and made it more relate-able. This time was no exception. Without recourse to a big fat Google blitz, I was forced to think through my topics more rigorously. For laughs, I took fewer cheap shots and told more stories.

What wasn't so heartwarming was my mood. I resented the fact that writing by hand took so damn *long*. And I was completely horrified by my own revision mania—a compulsion the delete key had been politely concealing all these years. It was nothing to use up an entire sheet of lined paper to produce a single useable sentence. I found that disturbing. Also confusing. Without the accustomed visual order of flawlessly justified chunks of prose, it was so much harder than I'd anticipated to maintain my momentum, let alone to preserve the logical flow of ideas. Even more of a struggle was trying to keep to my word length—a rigid but comforting constraint for any columnist, a sort of snugly fitting seat belt of the mind. On a Microsoft Word document, even without the word-count function, I could tell at a glance where I was within a few words, like an old woman who measures the flour for her loaf by the fistful. Writing in longhand I lost all sense of that inner compass. Between the cross-outs and scrawled interpolations, I never knew where the hell I was. Like a whiny kid in the backseat, it was a constant case of "Are we there yet? Are we there yet?"

Plus, it hurt. The last time I'd held a pen for that long was probably in second grade. Back then, my teacher described my handwriting as "lazy." It was swiftly apparent that the passing years hadn't done much to energize it. My words looked so provisional on the page, so vulnerable and unsure of themselves. I missed the visual authority of my favorite font (Century Schoolbook), the neat margins and perfect spacing: the hospital corners, if you will, on the page. But more than anything, I missed the sheer velocity of word processing.

I took touch-typing as an elective in eighth grade—it was that or sewing—and it was probably the single most useful course I ever did. (And I still believe sending my daughters' skirts out to a tailor for hemming was a small price to pay.) By the time electric typewriters came in, I was able to transcribe my thoughts almost as fast as they happened—often surprising myself in the process. "Good Lord, do I really think *that*?" I'd marvel, or cringe. The keyboard wasn't something I simply used for writing. It was almost a sensory apparatus.

I'd considered digging up an old IBM Selectric for use during The Disconnect, but quickly rejected the idea. Somehow, introducing a new gadget—however vintage—seemed to run contrary to the spirit of the enterprise. Even more horrifying, I could envision the whole family fighting for access to anything resembling a keyboard. No. It was better to stay pure and suffer the joint pain in silence.

After a few weeks of fighting the good fight, I slunk off with my laptop to a café, where I could have my technology and eat it, too. The familiar feeling of a keyboard under my fingertips was an almost sensual delight, it embarrasses me to admit. I could practically hear the endorphins whooshing through my tech-deprived neural pathways. McLuhan was right. A pencil is an extension of a finger writing in the sand. But our electronic media are extensions of our *brains*.

Once their initial excitement-slash-envious-incredulity died down, every adult who heard about The Experiment asked me the same question: "But what about their homework?"

Sometimes the tone was reproving, other times giddy with wonder. "They'll use the library of course," I'd reply smugly. Privately, though, I was worried. True, the miasma of multitasking through which the kids had been wading in recent years was hardly what you'd call optimal work conditions. Eliot's phrase "distracted from distraction by distraction" springs to mind. On the other hand, most of the time it did keep them at their desks with a homework window or two theoretically open. Some sort of assignment was generally under construction. And somehow or other it eventually got handed in. I very rarely policed anyone, and I'd certainly never been one of those parents who pitches in eagerly on "our" project on the vanishing dugong or the role of women at Botany Bay.

I didn't think then, and I still don't, that my kids' homework is really my business. Maybe it's a single parent thing, or maybe it's a writer thing, or maybe it's just a neglect thing. But whatever it is, it's worked—more or less—for us. (I've since found out ours is hardly an isolated case. The evidence shows that teens whose parents help them with homework are actually *less* successful in school, after all other variables have been controlled, according to a study of "Adolescents' Experience Doing Homework" published in 2008.[15])

In point of fact, there's not much evidence that homework is worth doing at all—by anybody. A recent study by University of Auckland researcher Professor John Hattie analyzed the effectiveness of 113 different teaching strategies. Homework straggled in at eighty-eighth. Hattie maintains he found "zero evidence" that homework helped to improve time-management skills, or indeed any other. He also observed that in the case of long-term projects, "All you're measuring is the parents' skills."[16] Vindication! In fact, a morbidly obese body of research has been showing pretty much the same thing for a long time. Homework makes no difference to primary-school kids at all; it helps bright children more than the less able; and busywork-style worksheets are functionally meaningless.[17]

I reminded inquirers I hadn't banned anyone from using computers, only from using them in our home. "Yeah, well, *home* is where you're supposed to do it," Sussy muttered. "Hello? That's why they call it *home*work." Be that as it may, as the weeks progressed each child worked out his or her own modus operandi.

Anni was the only one who approached the challenge with zest. She was actually looking forward to working on assignments at the college library. "I'm hoping it'll make me more organized," she told me as she searched for the car keys in her underwear drawer. A talented student, she'd fallen into the nasty habit of relying on her writing gifts to cover a multitude of rhetorical sins (she must have gotten that from her father) and habitually crossed her deadlines by a pug's nose. Like Proust, she wrote in bed and always had, ever since she'd gotten her first laptop in Year 8. Unlike Proust, she was also prone to spend entire evenings pimping her Facebook photos in there.

"It'll be good to come home and know I *can't* do any work, even if want to," she added. Although the separation of work from leisure was something I was looking forward to as well, I noted Anni's use of the word "can't" with foreboding. Whereas I struggled woman-fully to regain my mastery of paper and pen—and failed—the Digi-tal Natives didn't even go there. With the exception of Bill's and Sussy's math homework, it was as if they'd agreed that the whole concept of offscreen study was too outlandish even to try. I started to understand the point myself the day we went to uni to enroll Anni in her second-year courses and were told to go home and do it online.

"We *can't* do it online at home," I began, testily. I could sense Anni tensing up beside me. Maybe it was the fingernails drawing blood from my forearm. "And even if we could, why should we? We are here, in person and waving a checkbook. We don't *need* to be online . . ." Anni pulled me away, giving an apologetic shrug to the woman behind the counter, as if to say, "Birth trauma. There's nothing we can do, really." We had no alternative but to head to the library, on the other side of campus, and use one of the public computers there. LOL when it turned out they were all offline, and we were told once again to go home and try again.

Bill shuddered at the mere mention of the "L" word. "But your school has an amazing library!" I enthused. "Full of books and, and shelves and . . ."

"No way!" he cried, putting his hands before him fearfully as if fending off a rabid bat. He managed to do what he had to do at friends' houses, efficiently and without fuss, usually on the Sunday night before an assignment was due. I never asked to see his work, and he never volunteered to show it to me. But his school report at the end of semester two—a week or so after The Experiment con-cluded—corroborated my impression that he had continued to work at school to his usual dependable level, my doubts about what the lad was actually learning there (if anything) notwithstanding. "An A in English? What's *that* about?" jeered Sussy, reading over my shoulder.

"Stuffed if I know," came the answer.

But the truth was, Bill *had* been reading *To Kill a Mockingbird*. There'd been sightings. In fact, the boy who since Year 4 had subsisted on a literary diet of *Mad* magazines and the backs of cereal boxes—reading one book (the latest *Harry Potter*) every two years, whether he needed to or not—was reading more or less constantly. By the middle of month two, his "dry" time—the time not afloat in the water-polo pool—was divided between saxophone, listening to jazz, and ripping through novels like a buzzsaw.

Over the summer holidays, sheer boredom had driven our hero to a full-scale Rowling retrospective, reading all seven books swiftly and in sequence. When he put down *Harry Potter and the Deathly Hallows*, he looked as though he'd lost his best friend. "I finished," he announced listlessly, as he opened the fridge door for a ritual check, just in case some junk food had spontaneously generated in the last forty-five minutes. For the next couple of days, I waited almost breathlessly to see what would happen. Would he start all over again with *The Philosopher's Stone* and just keep cycling through for the next six months, stuck in a kind of Harry Potter purgatory? His friend came to the rescue with another series—something about wolves, or brothers, or maybe both. "Any good?" I ventured to ask as he neared the end of the fourth book. He shrugged. "I'm pretty sure I read them in Year Six, but yeah . . ."

I ended up tossing him the Haruki Murakami book a few weeks later. *Kafka on the Shore* is structurally complex, intellectually demanding, and vaguely spiritual, in a Tarantino-meets-the-Dancing-Wu-Li-Masters kind of way. It was an improbable read for a fifteen-year-old with a literary underbite. On the other hand, it *was* Japanese—like Pokémon and Naruto and so many other of Bill's pop-culture fixations. And it *was,* as of course I was soon reminded, full of jazz references. The next day, I asked him how it was going. "Good," he answered. "Weird."

"Promising," I thought to myself. That was sometime in early March.

Two months later, by Bill's sixteenth birthday in mid-May, he had stroked his way effortlessly across half of Murakami's considerable oeuvre. To celebrate, I bought him the complete works, or as complete as Perth's bookstores would permit. Even now, I have a hard time getting my head around that. For my son's sixteenth birthday, I bought him eleven books *and he was thrilled*.

According to a 2009 survey by the Consumer Electronics Association, 83 percent of U.S. teens believe technology helps them with schoolwork and learning—and only 23 percent reported that their parents restricted their use of technology.[18] Yet research shows the impact of media on our children's reading habits to be somewhere between negative and apocalyptic. News travels slowly, I guess.

But there are complications—and one of them is that our high-tech kids are not necessarily reading less, exactly. Nicholas Carr, writing in *The Atlantic* in 2008 on the question "Is Google Making Us Stupid?" and more recently in *The Shallows* (menacingly subtitled "What the Internet Is Doing to Our Brains"), points out that all our frantic Web browsing and text messaging probably means we are reading more today than a generation ago, when television was the alpha medium. Today, however, "It's a different kind of reading, and behind it lies a different kind of thinking—perhaps even a new sense of self"—as any parent who's ever sneaked a peek at her teenager's Instant Message conversation would agree.[19] ("Meh . . . WTF?! . . . Soz! . . . Bahhaha-hahahahahaha!") And one of the major differences is to do with depth, or the lack thereof.

Reading in the age of the Internet is skim-deep. It stays on the surface, which is why they call it "surfing." I prefer the more up-to-date term: "WILFing." Shorthand for "What Was I Looking For?"

WILFing refers to the habit of online free association that starts out with a specific purpose and ends up hours later . . . well, let's just say "elsewhere." It's not just kids who are afflicted by WILFing. Grown-ups are also at risk, even writerly types such as Nicholas Carr. "Immersing myself in a book or a lengthy article used to be easy," he observes. "My mind would get caught up in the narrative or the turns of the argument, and I'd spend hours strolling through long stretches of prose . . . Now my concentration often starts to drift after two or three pages. I get fidgety, lose the thread, begin looking for something else to do . . . Once I was a scuba diver in the sea of words. Now I zip along the surface like a guy on a Jet Ski."[20]

Maybe it's an occupational hazard to which journalists are particularly prone. I had certainly noticed the same symptoms myself over the past couple of years. What Harold Bloom has called the "difficult pleasure" of reading at full length—and in full depth—was becoming a rarity. Literally, I felt I was forgetting how.

I'd always had problems with long-term relationships. But come on. A novel? I couldn't commit to reading a whole novel? I'd put it down to hormones—those ever-obliging scapegoats—or maybe adult-onset ADHD.

And then The Experiment came along and forced me to take a closer look. And what I found was a very unpleasant family resemblance. Basically, I had developed exactly the same bad habits I criticized constantly in the kids—a sort of cognitive antsy-ness that took the form of boing-boing-boinging my way through a hundred sources and never settling down to digest any one of them. It seemed the menopause was *not* the message.

When I started working from an office again, as I did in mid-March, I had a desktop computer and regular Internet access for the first time since we'd pulled the plug. Being able to take notes on a keyboard felt illicit, and amazing. To write the story of Our Disconnect,

I'd already collected hundreds of articles and studies and dozens of actual books with actual covers—and all of them would need to go into the hopper. I couldn't wait to get started. Appropriately, one of the first cabs off the rank was *Distracted: The Erosion of Attention and the Coming Dark Age* by Maggie Jackson.

Although initially put off by Jackson's gloom—the subtitle captures the subclinically depressed feel of the thing—before long I was hooked. True, there weren't many laughs, or even much subtlety. (Prophecies of Armageddon are a bit like that.) But the book was so superbly researched, so scarily alive with statistics, examples, and anecdotes, that I found myself swept up in its argument almost against my will.

And yet . . . and yet.

At the same time as I was devouring *Distracted*, I found my attention wandering badly. I started in just doing my job: Googling various studies cited in the book. (Some didn't turn up on Google, so I used Factiva and ProQuest 5000 instead. Bingo!) Then I Googled Jackson herself. ("Ah, so *she's* the one who used to write that Balancing Acts column in *The Globe*!") Then I Google-imaged her. ("Nice hair. Wonder how old her kids are . . .") She mentioned Brueghel's painting of the fall of Icarus—I Google-imaged that too, ending up in a labyrinth of fascinating museum sites.

Several hours later, I found myself on Amazon.com, ordering a book called *Why He Didn't Call You Back: 1,000 Guys Reveal What They Really Thought About You After Your Date*. And that's when I came to. I snapped off the computer and reached for a stack of index cards—I obviously couldn't be trusted with a keyboard yet—and found my place in the book. "Nearly 45 percent of workplace interruptions are 'self-initiated,' " I read. Tell me about it.

Jackson argues—and I am clearly in no position to demur—that the informational acid rain falling everywhere in our culture, our own family rooms very much included, is eroding what she calls "the three pillars of attention": focus, judgment, and awareness.[21] We are captives of information—in the words of Walter Ong—dangerously adrift in an information chaos that means nothing and takes us nowhere. (See, I told you it wasn't fun.) But the worst part of all is: We think we're doing great. We think we're smarter and faster and more wised-up than any previous generation. Hey, it says so on Wikipedia!

Our children are supremely confident users of new media. (Their generation seems to think they invented the Internet, which is completely LOLworthy, when you stop to think about it.) My own kids are downright smug about their cyber-superiority, rolling their eyes at my leaden-fingered mouse technique, or making impatient grabs for the keyboard to show me shortcuts. Lord knows, they are right, too. Some of the time. When it comes to Web 2.0 mastery—social media, file-sharing, creating user-generated content, and the like—they are all over me like a bad case of acne. But for good ol', boring ol' info-crunching, there's nobody like a Digital Immigrant to show you a good time. All that onscreen "homework" they're doing, all that "I need the Internet for *research*, Mum!" notwithstanding. When it comes to actually learning anything online, it turns out the Natives really *are* revolting.

Study after study shows today's students "display a particularly narrow field of vision" when doing Internet research. They use "quick and dirty" ways of searching, "often opt for convenience over quality," and give up easily.[22] Pretty much the way they clean their bedrooms, in other words. Research shows that Generation M are more *confident* information processors than their elders. Their competence is another matter. A 2008 study of "adolescent Internet literacies" published by the International Association of School Librarianship observed kids using the Internet while they did homework and concluded that most needed a crash course in remedial Googling.

"Despite their extensive use of the Internet," the researchers found, "students lacked skills in many areas but particularly in locating information and critical evaluation of Internet sources."[23] The teens they studied universally began their searches with Google, "entering a very few keywords with no search markers." Of course, Wikipedia was usually the first site to appear. Students were aware that some sites were less reliable than others, but were unable to find a way around this. Most kids also believed that they knew more about the Internet than their teachers did and therefore felt resistant to instruction when it was forthcoming. Which it often wasn't—because teachers themselves had varying abilities. Parents only added to the confusion, researchers found. Many were fixated on the (highly exaggerated) risk of Internet predators. Others, obscurely, saw books and online content as opposing armies. They seemed convinced that education was about books "winning."

Overall—and this is both amusing and disturbing—the researchers found that none of the stakeholders—neither parents, teachers, nor kids—was particularly tech-savvy. They *all* had misconceptions, knowledge gaps, and skill shortages that made the late senator Ted "The Internet Is a Series of Tubes" Stevens* seem like an oracle.

Other research has found that teens enjoy their homework more when it is a "secondary activity" and socializing is the "primary activity." Astonishing, no? Equally duh-worthy is the finding that kids nevertheless do a better job when homework is the only thing they focus on. Their "affect" may be more "negative"—they may be pissed off, in other words—but their achievement levels couldn't be happier.[24] When it comes to homework, in other words, feeling good and doing good are entirely unrelated.

A recent survey found that 80 percent of teens reported that going a day without technology made them feel "bored," "grumpy," "sad,"

---

*The Republican Digital Immigrant.

and "uninformed." A week without technology was regarded as "severe punishment."[25] (Needless to say, I chose not to share this study with my children.) Our kids are happiest—or so they report—when they are plugged in. They are also laziest, least focused, and least productive. That does our heads in, because as modern, psychologically attuned parents dedicated to micromanaging the states of our kids' minds, at some level we assume that happiness is a prerequisite for achieving anything worthwhile.

As Sussy would say: as if.

Psychologist and father of three Michael Osit believes that feel-good technologies actively undermine the development of a basic work ethic, and are creating "a generation that is used to getting what it wants with minimal effort."[26]

"He says that like it's a bad thing," Anni objects. It's a valid point. Osit also rails against the corrupting influence of Japanese cuisine— "If the kids are eating sushi at age ten," he scolds, "what will they be asking for at fifteen and twenty-five?"—and suggests musicians who use applications such as composing software "Sibelius" are cheating.[27]

But beneath the conservative bluster, Osit is onto something critical. Our kids really do have problems distinguishing between work time and playtime. For them, as for the rest of us, the Internet is both playground and workstation, snack bar and kitchen garden. "Just imagine trying to stick to your diet if you were asked to spend all day in your favorite bakery shop, or trying to stop drinking while you worked as a bartender," writes Osit.[28] Keeping the boundaries in place—hell, even remembering that there *are* boundaries—is hard enough if you're a grown-up. For our kids, it may be impossible. Osit is a firm believer that it is a parent's role to take control of the media environment. Expecting kids to do so on their own is like Googling "parenthood" for a job description.

When we work at our very peak, it often feels like play. And when we play at our peak, it is often very hard work indeed. For young children, learning and playing are almost always indistinguishable—and ought to be. There is no *necessary* connection between fun and learning, but to argue that enjoying oneself somehow prevents learning is plain ignorant. Research shows that kids *do* acquire skills from all those hours spent socializing on their media. Which skills, exactly, nobody is completely certain.

A $50 million MacArthur Foundation study on digital and media learning surveyed more than eight hundred teenagers over a three-year period. "It may look as though kids are wasting a lot of time hanging out with new media," lead researcher Mizuko Ito told *The New York Times*, "but their participation is giving them the technological skills and literacy they need to succeed in the contemporary world."[29] Reading these words, I try hard not to think about Anni's obsession with Farmville—a time-sucking Facebook-based simulation game that involves planting fake crops and raising fake livestock in cooperation with fake neighbors. ("See what you can achieve!" my brilliant daughter said proudly as she showed me her two-dimensional "farm." It was hard to know whether to laugh, cry, or oink.)

"They're learning how to get along with others," Ito insists. "How to manage a public identity, how to create a home page."[30] How to harvest fake soybeans.

Neuroscientist Gary Small agrees that being online sharpens some cognitive abilities. Digital Natives respond more quickly to visual stimuli and are better at certain forms of attention—for instance, noticing images in their peripheral vision. They are also better skimmers, able to "sift through large amounts of information rapidly and decide what's important and what isn't," and they may have higher self-esteem as a result of a greater sense of personal autonomy and control. Finally, like other brain scientists, Small speculates about the evolution of neural circuitry "that is customized for rapid and incisive spurts of directed concentration."[31]

Media can be distracting while children do homework, but it can also pay unexpected dividends. In 2008, University of Phoenix researcher Selene Finch conducted an in-depth, qualitative, phenomenological analysis—aka, she watched a bunch of kids do their homework—and found that instant messaging was sometimes a useful learning tool, especially for seeking homework help from classmates. True, this "helpfulness" could also cross the line into full-scale cheating, as evidenced by the participant who admitted that she and her buddies would split up their assigned work, then IM each other the finished products. The girl told Finch she knew this "sounded unethical."

"We probably talked about homework online sometimes," one subject allowed. "My typing skills got really good," another enthused.[32] Overall, the benefits seemed rather weak.

A recent British study cited in *The Futurist* at the beginning of 2009 found that smartphones deployed in the classroom could be a "powerful learning tool," enabling students to set homework reminders, record lessons, access relevant websites, and transfer files between school and home.[33] A 2008 Harris Interactive/Telecommunications Trade Association survey found that 18 percent of teenagers believed their cell phones were having "a positive influence on their education," while 39 percent of teen smartphone users said they accessed the Internet on their phones "for national and world news."[34] (LOL!)

There is also a growing market in "homework management solutions" software such as Schoolwires Centricity, touted as an enabler of "teacher–student collaboration like never before." According to its website, among other features, Schoolwires "allows students to keep track of upcoming assignments" while enabling teachers to "deploy and manage multiple custom, interactive websites; to implement advanced Web 2.0/multimedia capabilities, such as blogs, podcasts, and photo galleries; and to enable all users—from novices to power users—to utilize the functionality that meets their needs and comfort levels." Yikes! Then there are Internet-based utilities such as Parent Portal, which allow parents to access their kids' grades and attendance, in addition to teachers' homework and lesson plans.

Schools in the tiny Arkansas town of Howe—where 75 percent of students qualify for subsidized lunches and the superintendent drives the bus—have had success using iPods loaded with lesson content, including "songs about multiplication facts" and teachers' notes transmitted directly from classroom-based SmartBoards.[35] Science teacher Jim Askew is proud that the federally funded program—it cost $1.5 million to build and maintain—means there are no books in his classroom. "I'll bet there's a whole bunch of science rooms around now that still have science books that say there are nine planets, but that's been changed since Pluto was demoted," he observes with pride. But Askew is equally adamant that using new media to enrich student learning is no quick fix, insisting, "Anybody who thinks technology saves teachers' time is wrong."[36]

It's a lesson many other schools have had to learn the old-fashioned way. Northfield Mount Hermon, for example, a private boarding school in western Massachusetts, pulled the plug on its laptop program when school officials realized more effort was going into repairing the computers than teaching or learning with them. At Liverpool High School in upstate New York, millions of dollars in government grants were awarded to provide laptops for all students in an effort to bridge the so-called digital divide between those who could afford home computers and those who couldn't. Seven years later, the school board president admitted, "There was literally no evidence" the laptop program "had any impact on student achievement—none." Schools in Broward County, Florida, leased six thousand laptops at a cost of $7.2 million. Here, at least, students reported two clear educational benefits: Their typing improved and they became astoundingly proficient at Super Mario Brothers. The district discontinued the program anyway.

More systematic studies of laptop learning send the same message. One such, which compared twenty-one middle schools that had laptop programs with twenty-one middle schools that didn't, found no difference whatsoever in student test scores. Maybe that's the good news. Because many teachers complain that "the box"—think Pandora and the ruin of man—actually impedes student learning. Listening to Sussy and her pals, it's easy to understand why.

Her school, which still has a laptop program, does what it can to block social media, IM utilities, and gaming on students' state-of-the-art MacBooks. But where there's a will, there's a way. And where there's Wi-Fi, there's an even faster way. When MySpace got blocked, everybody migrated to Facebook. When Facebook got blocked, they flitted away to Twitter. Not allowed to e-mail during class? Fine, we'll Skype instead. Cell phones banned? No worries. We can send SMSs cheaper using text4free.net.

The problem isn't confined to Year 10 schoolgirls awaiting hormonal updates. "People are going to lectures by some of the greatest minds, and they are doing their mail," MIT professor of the social studies of science and technology Sherry Turkle told *Time* magazine. "I tell them this is not a place for e-mail, it's not a place to do online searches. . . . It's not going to help if there are parallel discussions about how boring it is," she added. "You've got to get people to participate in the world as it is."[37] UCLA and the University of Virginia have given up appealing to students' better natures. They simply block Internet access during lectures.

Kids have never been more mentally agile—or more culturally clumsy—than the Digital Natives we are rearing and occasionally fearing. Anyhow, that's the opinion of Mark Bauerlein, author of *The Dumbest Generation: How the Digital Age Stupefies Young Americans and Jeopardizes Our Future*, who believes passionately that technology, instead of opening young minds to knowledge, "has contracted their horizons to themselves, to the social scene around them."[38]

Bauerlein argues that our kids' relentless exposure to screens has equipped them beautifully for more exposure to screens. And precious little else. On the contrary, "It conditions minds against quiet, concerted study, against imagination unassisted by visuals, against linear, sequential analysis of text, against an idle afternoon with a detective story."[39] Reading rates for young people suggest this is more than standard-brand conservative kvetchery. In 1982, two out of three eighteen- to twenty-four-year-olds were reading for pleasure. By 2002, fewer than 43 percent were. The number of seventeen-year-olds who "never or hardly ever" read for fun doubled between 1984 and 2004. Notes Bauerlein shrewdly, "If young adults abandoned a product in another consumer realm at the same rate, say, cell phone usage, the marketing departments at Sprint and Nokia would shudder"— and fix it, fast.[40]

And speaking of the sound of one hand doing homework, have you heard the one about the martial artist who meets the Zen master? "My swordsmanship is legendary throughout the land," the fighter boasts. "And what about your special powers? What can you do?"

The Zen master thinks about it and replies, "When I walk, I just walk. When I eat, I just eat. When I talk, I just talk."

And when he writes an essay on e. e. cummings, I'm betting, he just writes an essay on e. e. cummings.

Midterm Interview

April 4, 2009

Q: Well, guys, we're at the halfway point. Three months! Can you believe it?

BILL: It feels like longer.

ANNI: To be honest, I haven't actually noticed it that much. I'm surprised how little I've missed it. I thought it would just be killing me. But it's been fine.

SUSSY: The worst parts are there's never anything to do. I can't do schoolwork when I want. And I can't go for walks 'cause I don't have my iPod.

Q: Because it's impossible to walk without an iPod?

S: Yeah.

Q: You did some homework today, didn't you?

S: Yeah, we had to go to McDonald's to use the Wi-Fi but they didn't have an electrical outlet so we went to the X-Wray Café, but the backpackers stole the Internet . . .

Q: How do you steal the Internet?

S: Well, they downloaded all this stuff so it doesn't work anymore, so we went to the Angel Café . . . Yeah, it's hell nice in there. We stayed for three hours.

Q: Did you do homework all that time?

S: MySpaced for a bit, but mostly, yeah.

Q: Has The Experiment affected the quality of your schoolwork?

B: Nah.

A: Not really. It hasn't been a problem going to uni to get my work done.

Q: So all that stuff about "I need the Internet to do my homework" is . . .

A: A total cover!

Q: Do you feel like a different person in some ways?

B: I'm not a different person but some aspects of life have definitely changed. Basically, I'm playing sax more and reading more, yeah. But I think the technology thing was more of a trigger. Like, if everything went back on right now, I wouldn't change. Like, why would I? It's more fun than playing with the computer.

Q: Do you find you're listening to the radio more?

B: Nah, radio is crap. I listen to my CDs, and that usually makes me want to play my sax, so . . . yeah. Also, I think I play with Rupert and Hazel a bit more. But that's it.

Q: Would you say you spend more time thinking now?

A: Yeah. Like before I would spend heaps of time doing nothing, but it would take the form of maybe Facebook-stalking or something. Now, well, you find other ways to have fun. Going out more. Went through a big cooking phase for a while there, but that's died down now. Listening to the radio a lot more.

Q: What's that like?

A: Well, you miss the customization you have with your iPod. Like, you can't always have what you want when you want it—a song or a TV show or whatever. But you adjust.

S: I'm reading way more, and faster. I feel smarter. In the book section of MySpace most people are like, books? CBF!

Q: Yeah, but you've always been a reader.

S: Yeah, but I'm reading more intense books now, and bigger ones. Not just like *Princess Diaries*, LOL! Well, I'm still reading them too but . . . Like *Prep*. I started that a million times but never finished and now I have. And *Bright Shiny Morning*, which was really, really good but some of the facts were, like, hell boring . . . and now I'm reading *Finding Alaska*, which I just started tonight, about some guy at a boarding school. Oh, and that David Sedaris one—can I lend that to Sean, by the way? I think he would really like it. Oh, and PS, this weekend Fia's having a partay . . .

Q: Anything else?

s: Done more eating, for shiz . . . But yeah, I'm so bored all the time. I want an iPod!

q: How have your friends reacted?

s: First they go, "WTF?!" Then, "Cool! Your mum's a writer?" LOL. Georgie was like, "I don't care. We can play the Hannah Montana Trivia Game," and Ali was like, "No computers? Cool, that's even better." Lil's always up for a good board game . . .

b: They all say it sucks when I tell them about it. Like, that sounds really inconvenient and your mother is really a freak and stuff like that. Just like, "Why would she do that?"

a: Some are, "Oh God, I can't believe your mum is making you do that," but most people are just, "Oh really? Fair enough." Actually adults have a more extreme reaction than kids do.

q: You're kidding.

a: No, really. Every one of your friends, they're like, "Why can't I get hold of your mum?" and then I tell them and they're like, "OOOOHHHH MYYYYYYY GODDDDDD!!! I CAN'T BELIEVE THAT!!!"

q: Interesting!

a: Yeah, they're like, "What about your homework?!"

Loss of Facebook: Friending the Old-fashioned Way

Anyone who thinks they have 200 friends has got no friends.
—RAY PAHL, *professor of sociology, University of Essex*[1]

Week eight of The Experiment. It's high summer in Western Australia—hot, dry, and as dull as an assistant principal on a first date. We love our city (most of the time). It's so clean. So safe. So pretty. But it comes honestly by its nickname: Dullsville. Perth is a place where stores still close at 5:30 p.m. during the week, and all day on Sundays. Where restaurants that serve meals after 9:00 p.m. are as rare as a kookaburra's canines, and nightlife as we know it—unless we happen to be a marsupial—is unknown.

For preschoolers and old-age pensioners, Perth is probably as close to paradise as you can get without a doctor's prescription. But for the rest of us, life can go just a teensy bit slow mo'. Teenagers—with their high need for social stimulation—suffer especially from all this freaking serenity. So I guess it's not surprising that binge-drinking starts early in these parts, or that vandalism and petty acts of public violence are more common than in many much bigger cities. The average teen has sex at the age of fifteen here. ("Thank God, my kids are above average," I think every time I read that statistic.)

In some ways, or so you could argue, Perth kids need their media more than most, what with the tyranny of distance they experience day to day simply by virtue of living so literally on the edge of things. That's something I worried about a lot in the early days of The Experiment. I was worrying about it that very midsummer night, driving home from a concert through the eerie stillness of a Saturday night in the world's most isolated city.

And then, pulling into the driveway, I hear funny sounds coming from the living room. And voices. Loud voices. Loud *male* voices. My heart lurches in my chest. I don't have a cell phone anymore, so there's no way anybody can contact me while I'm out. Up to now, I've been fine with that. In fact, I've been ecstatic with that. But at this moment . . . ? I race to the open front door and that's when I see it. I stand there in shock, my mouth as round as a laser disc.

It's a bunch of kids, five of them, around the piano.

They.

Are.

Singing.

Toto? I have a feeling we're not in Kansas anymore.

"What's next on the agenda, dudes? A taffy pull?" is what I'm thinking but don't dare say. If they are sleepwalking in another decade, far be it from me to disturb them. This, I realize, as I practically tiptoe to my bedroom, strenuously feigning nonchalance, is the moment I've been waiting for. Doing homework, sure. Reading and listening to music, absolutely. Practicing saxophone, cooking meals, sleeping and eating better—all of that has been extremely gratifying. At times verging on the magical, even. But it's this above all else—this, what would you call it? Connecting? One to the other, in real time and space, in three dimensions, and with all five senses ablaze. . . .

I realize they are a long way from roasting a woodchuck over the open coals. But I'm having a Thoreauvian moment, nonetheless. I get

into bed and rifle through *Walden*. I'd give anything to hit CtrlF—the
"Find and Replace" shortcut—right now. The irony of that is not lost
on me. But neither, as it happens, is the passage I've been seeking:

"I wanted to live deep and suck out all the marrow of life, to live so
sturdily and Spartan-like as to rout all that was not life, to cut a broad
swatch and shave close, to drive life into a corner and reduce it to its lowest terms, and, if it proved to be mean, why then to get the whole and genuine meanness of it or, if it were sublime, to know it by experience . . ."

I underline it in green, to the strains of Lady Gaga's "Pokerface."
It's awful. And I've never heard anything so lovely.

The next morning I read the passage to Sussy. "Do you understand
what Thoreau is getting at, honey?" I ask. "I think so," she replies.
"It's like . . . RL, right?"

Midterm Interview cont'd

Q: Have there been any positives for you so far?
ANNI: I think we're closer as a family, for sure.
Q: Why's that?
A: Because we talk more. It's like, "There's people in this house . . .
   Let's talk to them!" Suss and Bill come into my room now. It's
   been years since they've done that—just to hang out and have
   conversations. Just to talk, you know?
Q: Has The Experiment changed your relationships at all?
SUSSY: Anni and I are like we used to be. We're tight again.
Q: Is that out of desperation, or . . .
S: Probs! (Laughs)
Q: Seriously, how's it different?
S: It's like, we chillax. We tell each other stuff now, like we used to.
   She helps me. I help her. We play the dice game. We play our
   creepy little ring game. . . .

Q: What about Bill? How's your relationship with him changed?
S: Um, I want to kill him more, because of the sax. It's just. Soooo.
     LOUD!

> We've got nothing against the Internet, but when people are surf-
> ing the Web, they're missing the best part of life—being together!
> That's why we created the first Web site devoted to helping people
> spend less time online and more time with each other. For starters,
> we've allocated just enough time to browse every link, but not a
> second more. So enjoy your three minutes, then get out there and
> make face time. Chop, chop. Time starts now.
>
> —"Make Facetime" promotion, Dentyne.com

A website devoted to helping people spend less time online? Well, I
guess I've heard of stranger things. Like that after-school show where
the hosts are constantly urging kids to get outside and ride their bikes.
Seriously, it makes Huxley's *Brave New World* look like a press release
for Pfizer.

I, too, wanted to help people spend less time online, just like the
folks who brought you minty-fresh breath. If only I'd thought of creating a website instead!

When McCann Erickson created the "Make Facetime" campaign
for Dentyne brand-owner Cadbury in September 2008, they were
aiming straight for the kisser. The idea that the ads could induce
under-twenties to swap their technology for a stick of gum and a good
old-fashioned chin-wag was always going to be hard to swallow . . . not
to mention impossible to digest. ("I think most college kids would roll
their eyes," commented one sociologist drily.) But the fact that it was
tried at all is interesting. So is the site itself, which includes a couple
of not-entirely-user-friendly social-networking utilities—well, they
stumped this Digital Immigrant—and something called the Smiley

Chamber of Doom, which shows animated emoticons being maimed and tortured. (I'm down with that.) Oh, and it really does cut out after three minutes. Which is kind of amazing and also kind of annoying—especially if you happen to be taking notes.

Closer to home, Dôme Coffee—the Australian-owned café chain—is running a series of magazine ads with a strikingly similar anti-tech/pro-talk theme. "One friend face to face beats 100 on Facebook," reads a Confucius-like headline on a recent full-page ad. It features a photo of a cluttered lunch table and two broadly smiling women friends in their thirties . . . *staring at mobile phone screens.* (I show the ad to Suss. "See anything wrong with this picture?"

"Is it something about feminism?" she asks, warily.)

The information paradox—that the more data we have, the stupider we become—has a social corollary, too: that the more frantically we connect, one to another, the more disconnected our relationships become. We live in an age of frenzied "social networking," where up to a quarter of us say we have no close confidante, where we are *less* likely than ever before to socialize with friends or family, where our social skills and even our capacity to experience empathy are undergoing documentable erosion.

Our, quote-unquote, family rooms are docking stations now. We have five hundred or six hundred "friends," and no idea who our next-door neighbors are. We affiliate with "communities" based on trivia—a mutual appreciation of bacon, a shared distaste for slow walkers. And doing so in a spirit of heavy-handed irony hardly ennobles the enterprise. We have sold out social depth for social breadth and interactive quality for interactive quantity to become what playwright Richard Foreman calls "pancake people": "spread wide and thin as we connect with that vast network of information accessed by the mere touch of a button."[2]

Or at least that's one side of the argument. There are others who argue that our social connectivity is not fraying at all, but simply undergoing some much-needed rewiring. They point to the growth of online communities—from social-networking utilities such as the fascinatingly telegraphic Twitter, to the entire virtual worlds of Second Life and World of Warcraft. They show how new media are bringing families together with instantaneous digital contact via text, sound, image—or all three at once. ("Have you Skyped Grammy and Grandpa to say thank you for that birthday money yet?") They remind us that for Digital Natives, time spent grooming one's online relationships on Facebook or Twitter alone amounts to a sizable part-time job. That the happy confluence of wireless Internet and portable media means they are *never* alone, never out of touch. "Only connect" is what these people *do*.

So, are we really more connected and less alone than ever before? Perhaps the truth lies somewhere in the middle. But I don't think so—and maybe that's why it's all so confusing. My own observations suggest that the truth lies at both extremes. "Information explosions blow things up," remember? In this case, the land mine seems to have taken out the Via Media (literally, the middle way) altogether.

We are both much, much better connected, *and* in clear and present danger of forgetting how to relate. Well, I guess that's why they call it a paradox.

Watching as my kids adjusted to the aftershocks of life without social media drove the point home again and again. Maybe a night on Facebook really has become the moral equivalent of standing around the piano singing show tunes. But while the quality of each experience, and the skills and habits they call into play, are both certifiably "social," they also happen to be certifiably antithetical. Messaging, poking, posting, uploading, and gifting faux livestock to your friends can be absorbing, entertaining, even challenging. But "getting together" it ain't. I knew that before, of course. But The Experiment just sort of turned up the volume—not just for me, but for all of us.

The impact on our relationships as a family was even more dramatic, as we found ourselves "tuning in" to one another in unexpected ways. We lingered more around the dinner table—and talked. We watched the fire together—and talked. We pulled out old photo albums—and talked. We played board games—and talked. We climbed into one another's beds and read the paper—and talked. Are you getting my drift? Realizing that, to quote Anni, "There are people here. Let's talk to them!" came as an epiphany to all of us, I think. For me, that provoked guilt and delight in almost equal measure. But hey . . . isn't that what being a parent is all about?

Conversation, studies show, is good for the brain. "No, duh!" as Sussy would say. But these days, it seems, some of us need convincing. According to UCLA neuroscientist Gary Small, talking to people face-to-face—as opposed to face-to-Facebook—provides "greater stimulation for our neural circuitry than mentally stimulating yet more passive activities," including reading.[3] A 2008 study found that subjects who'd spent ten minutes chatting with friends scored better on memory tests than those who'd spent the same amount of time watching TV or reading a book, and the same went for those who'd engaged in "intellectual activities," in this case, solving puzzles.[4] Think about it. More time spent in face-to-face conversations *could* mean your child remembers where he left his laptop charger.

Online chatting, on the other hand, has been linked to symptoms of loneliness, confusion, anxiety, depression, fatigue, and addiction. Says Small, "The anonymous and isolated nature of online communication does not provide the feedback that reinforces direct human interaction."[5] A study published in the journal *CyberPsychology and Behavior* found that shy people spent significantly more time on Facebook than more outgoing individuals—although they had fewer "friends"—and enjoyed it more too. The possibility of "a reliance of shy individuals on online communication tools" concerned the researchers, psychologists at the University of Windsor in Ontario, Canada.[6]

Half a world away, Japanese psychologist Tamaki Saito has coined the term *hikikomori* to describe a new breed of young social isolates. The Japanese Ministry of Health defines *hikikomori* as "individuals [80 percent are estimated to be male] who refuse to leave their parents' house, and isolate themselves away from society and family in a single room for a period exceeding six months." But the definition leaves out an important fact: *Hikikomori* are often, paradoxically, the most "connected" individuals in Japanese society.

Many *hikikomori* sleep by day and spend their nights watching manga, gaming, and surfing the Net, surfacing only to sneak into the kitchen for food while the family sleeps. An entire industry has sprung up to address the phenomenon—from parent support groups to online counseling services—but the epidemic continues to rage. In July 2009, Osaka police attributed to *hikikomori* a spate of street attacks by "apparently troubled people venting their frustration on total strangers."[7] One young man admitted he didn't care who he killed. "I'd grown tired of life," was his only defense. Experts believe *hikikomori* may turn to violence because of their lack of social skills. "Once they come to be considered weird, they prefer to be alone rather than feeling awkward among other people," explains Toyama University academic Yasuhiko Higuchi. "They then commit an extreme crime after magnifying their stressful thoughts and having no one to talk to."[8] Many *hikikomori* have failed to form a proper relationship with their parents, he adds. Uh-oh.

Pre-Experiment, I mention the term *hikikomori* casually to Bill. He looks up briefly from his game—where a strapping youth is beating the cyber-crap out of a hulking avatar with incongruously girlish hair—and looks down again. "You're pronouncing it wrong," he mutters.

Turned out Bill knew all about *hikikomori*. In fact, he'd watched an anime series about them. "Really? Where'd you get that from?" I asked. "Um, *Japan*," came the reply, the "duh" unspoken but implied. He'd downloaded the show from a file-sharing site. "So, what do you think of them, then?" I asked, in that annoying faux naive manner beloved of therapists, parole officers, and mothers.

"I think they're cool," he replied evenly. (Like most fifteen-year-old boys, he could spot a cautionary tale at twenty paces.)

"Are you kidding?" I sputtered. "They're *mentally ill*! They have *no life*!"

He looked up once more, between body blows. Maybe he didn't *say*, "It takes one to know one," but it was there in his eyes.

I was still enjoying intermittent eye contact with my children (although Sussy, aka Thunder Thumbs, was developing an alarming facility for texting while doing just about anything: talking, eating, walking, and more than once, I swear, during REM sleep) but you didn't need to be a detective with the Osaka police force to notice that our opportunities for sustained sharing had become increasingly nasty, short, and brutish. Their online world had become "the point"—of existence, I mean—and every other kind of interaction constituted a tangent. An interruption. I was conscious of how often I approached them with words like, "Can you just pause that for a moment and . . ." or, "After you sign out, would you . . ." or, "I don't need you to log off, but . . ." It was as if life, real life, were a game they'd lost interest in after the first couple of levels.

Looking around our family room, at the children sitting frozen at their screens, I would be reminded of Swiss architect Max Frisch's definition of technology: "The knack of so arranging the world that we don't have to experience it."

Relating socially, whether one to one or in groups, seems so fundamental to human nature. The notion that we might need to practice these skills—to practice being human, really—seems odd to me, and perhaps to you too. But neuroscientific evidence reminds us that the pathways in the brain that facilitate interpersonal skills, empathy, and sound social instincts are created, not born. In the case of individuals "who have been raised on technology, these interpersonal neural pathways are often left unstimulated and underdeveloped," observes one expert.[9] Despite their higher IQs and bulging thumb muscles, in other words, The Young and the Listless do show deficits in basic social skills such as empathic listening, and interpreting and responding to nonverbal cues in conversation.

Some observers have gone so far as to suggest technology may be driving us all toward a kind of social autism—wrapped safely but suffocatingly in our digital bubble wrap, uninterested in and/or threatened by the world outside, and supremely ill equipped to deal with it. Alarmist though it may sound, it's not entirely far-fetched. In fact, recent research suggests there may even be a link between chronic technology use and *clinical* autism.

Whatever the cause, autism rates have skyrocketed during the digital age. Today, according to figures from the European Union Disabilities Commission, autism afflicts one in every fifty-eight children—an increase of up to 500 percent since records have been kept. Many theories have been advanced to explain the epidemic; almost all have been disproved. One that has not is the theory that started out as a parent's hunch.

Michael Waldman, an economist at Cornell University, was devastated when his two-year-old son was diagnosed with autism spectrum disorder. But he was also skeptical. He'd noticed that since the birth of their second child a few months earlier, his son had been spending more and more time watching television. Privately, he wondered whether the boy's socially phobic behavior was not a "disorder" at all, but simply an aggravated case of tuning in and . . . well, tuning out.

Waldman placed restrictions on the child's media habits and had him retested. When his "condition" improved and then disappeared entirely, it seemed like a miracle. But economists, thankfully, don't believe in miracles. Waldman cast about for a way to study his hunch about a link between autism and television viewing. And finally the answer came to him: rainfall data. Stay with me on this one.

Waldman reasoned that kids in rainier climates watch more TV— which is true, by the way—and therefore that regions with higher-than-average precipitation might also feature higher-than-average rates of autism spectrum disorder. He compared California, Oregon, and Washington—the rainiest states in the United States—with the rest of the nation, and he found his answer. There *was* more autism in these states. He then looked at only those families who had cable TV subscriptions in these high-precipitation regions, and the correlation was higher still.[10]

When Waldman's study was published in the November 2008 issue of the prestigious *Archives of Pediatrics & Adolescent Medicine*, it provoked a perfect thunderstorm of abuse and criticism. But the data remain standing. A 2009 article in the *Journal of Environmental Health* concedes Waldman's point about the link between rainfall and autism rates, but is more equivocal about causes. Perhaps the real culprit was not TV at all, but vitamin D deficiency, or increased exposure to household cleaners?[11]

No one, certainly not Waldman, would argue that television or any other medium "causes" autism, a dizzyingly complex disorder involving sensory, motor, and cognitive difficulties, as well as social ones. But the possibility that chronic media use may act as an environmental trigger for kids with an underlying genetic vulnerability is being taken seriously indeed.

But we don't need to drag autism in, kicking and screaming, to explain our kids' empathy deficits. Let's not forget, narcissism comes naturally to teenagers. There's even a specific region of the teen brain that controls their tendency toward selfishness. Digital Immigrants make use of the prefrontal cortex when considering how their decisions will affect others. Natives use their temporal lobes, which are slower and less efficient. Their underdeveloped frontal lobes make teens feel invincible ("Pregnant? Me? As if!"). At the same time, they ensure impaired judgment about almost everything: from how to choose a phone plan to how to choose a boyfriend. Our kids won't always be this clueless, neuroscientists promise. In theory at least, later brain development will enable them to delay gratification, to accurately assess risk, and—eventually—to consider the feelings of others.

In other words, we can't blame our kids' digital distractions for *all* their ditziness. Nor is it necessarily true that kids *either* spend huge amounts of time with media *or* they engage in lots of nonmediated activities. The Kaiser Family Foundation's Generation M study found just the opposite, in fact. Contrary to researchers' expectations, it turned out that "heavy overall media users also tend to spend more time engaged in several non-media activities than do light and moderate media users." Specifically, the 20 percent of eight- to eighteen-year-olds who were the biggest self-reported screen freaks were also the ones who spent the most time "hanging out with parents, exercising, and participating in other activities such as clubs, music, art, or hobbies."[12] Interesting. Especially given that the *average* amount of time kids spent on-screen in that study was 8.5 hours. You've got to wonder: When were the heavy users actually doing all that art and music and stamp collecting? In their sleep?

At the age of fourteen, Joan of Arc was leading the French army to victory in the Hundred Years' War. Sussy, also fourteen, struggles to change a fitted sheet. Eighteen-year-old Anni can be heard whimpering when she discovers the can of baked beans she wanted for lunch has no ring-pull. And my son the electronics whiz, who's been putting together robots since he was eleven, claims he hasn't quite gotten the hang of the dishwasher yet. Honestly. Who do these people think they are—somebody's husband? And, more to the point, how did they get that way? Personally, I blame the guy who invented those Velcro shoe-fastener strips.

Is it just in our household that teenagers struggle with skills and competencies that were once taken for granted by the smallest children? Evidently not. Some observers have suggested that Digital Natives—aka the Pull-Ups Generation—may be suffering from a kind of global life-passivity that goes way beyond garden-variety teen cluelessness. While acknowledging the universal truth that older generations inevitably view the younger ones snapping at their heels as degenerate, unmannerly, and incompetent—the technical term for this being "envy"—there does seem to be something new and scary going on here. In the United States, colleges have introduced undergraduate courses in basic life skills such as banking and doing laundry and ordering from a restaurant menu. (Remedial can-opening, anybody?)

Twenty-five years ago, critic Neil Postman argued that the rise of the global village would spell the disappearance of childhood. Among today's iGeneration, it has arguably elongated toddlerhood. After the equivalent of a full working day in front of their screens, is it any wonder our children have little patience for practicing life and all its funny little ways?

When two little girls got trapped in a storm drain near Adelaide in September 2009, they might never have made it out alive. Thank heavens the ten-year-old and twelve-year-old both had cell phones, and, like all Digital Natives, they knew exactly what to do with them.

They updated their Facebook status, of course.

Miraculously—but then again maybe not so miraculously—a school friend was online at the time and contacted the emergency services.[13]

Five years ago, social networking was something you did over drinks on a Friday night—and the only people who had five hundred friends were first-division lottery winners.

Today, thanks to the social media utilities Facebook, MySpace, and Twitter, only freaks and losers and people's mothers (if that's not a redundancy) are satisfied with having a few close friends. For everybody else, apparently, friendship—or, more accurately, "friending"—is the new Versace, a form of conspicuous consumption tailor-made for a GFC-shaken world.

In "the black and white days," we used to think communications technology was all about . . . um, communicating. As quickly and as efficiently as possible. You were trapped in a storm drain. You rang triple-0. You wanted to go on a date with somebody. You called and asked them out. You liked somebody's music. You bought their album. Today, that kind of no-frills approach seems so naive, so lacking in style and nuance and suspense. Calling somebody because you have a question to ask—or a life to save—is like being hungry and eating meatloaf. There's no art to it.

At the English restaurant The Fat Duck (recently judged the second-best eatery in the entire world), you can order bacon-and-egg ice cream, or lime and green-tea meringues poached in liquid nitrogen. They might not sate your hunger, but at this level of fine dining, hunger itself is not the point. In fact, it's a little uncool. The key to truly world-class dining out is all about the disconnect between food and hunger. Need, in other words, is not where it's at.

Well, Facebook is like that too. It's at least as much a performance medium as it is a communication medium—a stage on which to enact, perfect, and publicize "you" (whoever the hell that is). Asking and answering questions, eliciting or exchanging information—these things do happen. But on Facebook and other social media, including text messaging, they happen indirectly, unfolding in sideways steps like an origami flower or an art house film.

"Going out tonight?" I ask Anni on Friday afternoon. "Probs," she replies. "I just messaged Alex to say I'd message him later." A couple of hours later—and keep in mind I'm only trying to figure out what to cook for dinner—I try again. "Message Alex yet?"

"Nah, he messaged me before to say he'd message me later."

I decide to go ahead with a family meal, but by the time it's served, Anni tells me she couldn't eat another bite. She's been snacking on soy crisps for the last couple of hours, and anyway, she reminds me, she's probably going out for dinner. "But it's seven-thirty!" I sputter. "Surely you know whether you're going out for dinner by *now*."

She looks at me with a mixture of pity and disgust. "Why would I? I'm not even hungry." She glances back at her screen, where a new message has landed with a satisfying thud, and snickers gleefully. "Plus, I've just messaged Holly to say I'll meet her in the city later."

"Oh, okay. Well, what time are you meeting her?"

"Dunno. I said I'll message her from the train."

I take a cleansing breath. I promise myself I'll just leave it at that. I break my promise.

"Can you just tell me why it's necessary to leave everything till the last possible minute?" I genuinely want to know.

"Can you just tell me why it's necessary to be such a control freak?" she asks. I suspect she genuinely wants to know too.

"Diffuse" is a nice word for this style of communicating. Other options include "confused," "disorganized," and "utterly lacking in focus." Facebook status updates ("Still in storm drain! LOL!") or tweets are even less directional. There is no targeted recipient at all. Like a smoke signal or a billboard, these messages are broadcast indiscriminately. It's not a case of me talking to you, but of me talking to whomever in my community is online and paying attention. You don't address the envelope. You simply "put it out there," as Sussy would say. ("Mum, do you realize I've never, ever been to Paris," she announced at dinner last night, apropos of absolutely nothing. "I'm just putting it out there.") People don't reply, exactly. They "comment." They might say "I like this" with a little thumbs-up icon—presumably not if you're in a storm drain though—or throw a strawberry at you, or some other bon mot.

Information isn't the only commodity that becomes more diffuse on Facebook. Friendship itself does too, insist some observers.

At the time we unplugged, the average number of "friends" in a Facebook network was 120, according to Facebook's in-house sociologist Cameron Marlow.[14] Today, the figure has risen slightly to 130. Rather unsurprisingly, women tend to have more friends than men. In Anni and Sussy's age group having fewer than two hundred or three hundred is a sign of social backwardness, though "Not for guys," Anni tells me. Five hundred friends or more is nothing special. In an article titled "You Were Cuter on Facebook," even teen-focused *Cleo* magazine warns, "We are choosing quantity over quality." To illustrate the point, writer Bessie Recep recounts the tale of a friend presently documenting an eight-week European holiday at the rate of 350 digital images a day. Surely even the strongest friendship would stagger under the weight of viewing 20,000 holiday snaps, Recep muses. "I don't even want to think about how much time that'll take (time that could be spent creating my own life experiences and not just reliving someone else's)."[15]

Friends and photos have a lot in common in the digital age. There's no end to the number you can have—but just try to find the good ones when you need 'em . . .

Oxford University anthropologist Robin Dunbar, an expert in

social networking in humans and other primates, agrees with me—that one's contact list, while in theory infinite, is in practice subject to some pretty rigid restraints. Our capacity for "friending" is not only finite, but predictably finite. In fact, it's reducible to a number. Dunbar sees Facebook as a form of social grooming, exactly like that done in the wild by our ape cousins. The reach of any individual network—whether of people we "comment" or pick lice off—is strictly limited by its species' cognitive power. For human primates, Dunbar has calculated the number at around 150. Researchers now refer to it as the Dunbar number, and it has been found to be relevant across a wide range of human groups, from corporate divisions to Neolithic villages to Facebook networks.

Cameron Marlow's findings for Facebook suggest our core network capacity—the people we interact with specifically and reciprocally—is much smaller still. An average Facebook user with 120 friends—me, say—will generally communicate (in the old sense of the term) with only seven of them. That's about 6 percent. The other 94 percent are pretty much there for show. But the Facebook user with five hundred friends—Sussy, say—will only directly interact with about sixteen or a mere 3 percent of the pool. "Networking" is almost certainly a misnomer for all this. What Facebook users are really doing is "broadcasting their lives to an outer tier of acquaintances who aren't necessarily inside the Dunbar circle," notes Lee Rainie, director of the Pew Internet & American Life Project. "Humans may be advertising themselves more efficiently," another expert concludes, "but they still have the same small circles of intimacy as ever."[16]

In a social landscape dominated by "friending"—a gerund that still sets Digital Immigrants' dentures on edge—the word "friend" has arguably lost more value than the Vietnamese ng (a currency valued at around one 1,800th of a U.S. dollar). "When introducing a real friend to a new acquaintance, I often feel the need to call my friend 'a dear friend' or a 'close friend,'" writes University of Toronto's Neil Seeman. "'Friend' requires an adjective these days, since it otherwise feels empty. We've dumbed adult friendships down."[17] Only four-year-olds call everybody who says hello to them a "friend." But suddenly grown-up people who ought to know better are doing exactly that, carrying on like Casper the Friendly Ghost or Sniffles the Mouse (who, if memory serves, once tried to make friends with an acorn). I recently rejigged my own Facebook account to create two lists: "Actual Friends" and "Acquaintances at Best." The latter seemed more diplomatic than "Total Strangers."

As of June 2010, Facebook had 400 million monthly active users worldwide. Between 2008 and 2009, membership had doubled in the United States alone—where, just for the record, 38 percent of the total population has Facebook accounts as of this writing. Yet 70 percent of users are outside the United States. Worldwide, we currently spend over 500 billion minutes per month on Facebook.[18] That's roughly an hour and a quarter for every man, woman, and child on the planet.

Legend (and now a major motion picture, no less) has it that Facebook was founded by Mark Zuckerberg in a Harvard dorm room in 2004. Two years later, Bill Gates paid $249 million for a 1.6 percent share. As Clive Thompson observed in 2008 in *The New York Times Magazine*, Facebook's greatest innovation—and what makes it unique among other social utilities—is the News Feed: the very useful engine that broadcasts changes in a user's page to everyone on his or her friend list. Like many other users, I was aghast the first time I discovered how it worked, which I did when the humiliating news "Susan Maushart has updated her birthdate!" was flashed around the world. (I was only correcting a typo, I swear!)

The effect of News Feed is not unlike a social gazette from the eighteenth century, or, in Thompson's words, "like a giant, open party filled with everyone you know" except you're able to eavesdrop perfectly, on everyone, all the time.

When Zuckerberg trialed News Feed, early subscribers freaked out. But after a few days the tide of protest subsided. Within a few weeks it had given way to a landslide of support, and subscribers. Through the magic of News Feed, Facebook users could now enjoy minute-by-minute updates detailing the most trivial details of their friends' deeply humdrum lives—a gossip column, if you will, for nobodies.

The microblogging site Twitter, as even the least assimilated Digital Immigrant must by now be aware, works in much the same way, via posts—i.e., text messages—of not more than 140 words that answer the question, "What are you doing right now?" Powerful Twitterers—celebrities, politicians, and journalists in the main—who lead more tweet-worthy lives, use the site to broadcast everything from red-carpet gossip to fun-sized musings on foreign policy. The world's top Twits command millions of followers, and the Twitterverse is—as they say in the classics—expanding. Digital Immigrants who demand to know what's the point of Twitter (as so many of us do) simply show their age-slash-cluelessness. Often, there is no point, at least not in the Gutenbergian sense that communication is about the useful exchange of information. (In this respect, the term "social utility" is almost hilariously inaccurate.) Twitter describes itself as a "global cocktail party thrown by regular people." Does a party have a point? For that matter, does a cocktail party with no alcohol? But I digress.

Thompson and other observers argue that what social media such as Facebook and Twitter deliver is simply contact itself—"ambient intimacy" or "ambient awareness." Explains Thompson, "Each little update—each individual bit of social information—is insignificant on its own, even supremely mundane. But taken together, over time, the little snippets coalesce into a surprisingly sophisticated portrait . . . like thousands of dots making a pointillist painting."[19] It's not friendship exactly, Thompson concedes. It's more like artwork. Reading over my own sadly neglected Facebook page, it seems closer to a really crappy craft activity. Less pointillist painting than pipe cleaners on an empty egg carton.

The emerging etiquette of "friend requests" suggests there are plenty of nuances yet to be explored. Parents friending their own kids, for example—a practice Bill describes to me as "disturbed and barbaric." Barbaric? "In the same sense as the live sheep trade is barbaric," he explains. "Because it causes pointless suffering."

Sussy agrees, and Anni does too, largely. Before The Experiment, she and I became "friends," but I was careful to respect the boundaries. (No gratuitous photo-commenting, no hectoring wall posts, no snide comments about Farmville.) Yet the stigma seems to skip a generation. When I created a Facebook account for my mother and sent friend requests to her grandchildren, all six accepted immediately. "Does it bother you that Grammy checks out your photo albums?" I asked Sussy.

"I am semi-freaking out," she admitted.

But I guess that's what families are all about, right?

When Bill "added" his favorite teacher last year, it was my turn to semi-freak out. But a study reported in *Psychology Today* of university-based Facebook users found that academics who disclosed information about their social lives on their profiles created a more comfortable classroom climate and increased student motivation.[20] Yeah, but to do what? On the other hand, a third of students surveyed believed their teachers should not be permitted access to Facebook at all, citing privacy and "identity management" concerns.

An article I came across in a law journal examines the advisability or otherwise of legal professionals friending witnesses.[21] *The Wall Street Journal* reports that U.S. tax agents have joined in the fun as well, using social networks to friend and apprehend suspected tax cheats.[22] The tendency of networking media to redraw traditional social boundaries—whether between generations, or school cliques, or authority figures and subordinates—is part of the attraction. We are all equal in the eyes of Facebook, or among those whom we tweet. Or so, at least, runs the mythology.

"Twitter is all about stalking celebrities," Sussy informs me confidently during month four. "It's like, you add them and then you know everything they do." She doesn't have an account yet, of course. Weirdly enough, I was the one who explained to her what Twitter was. But in the weeks since we embarked on The Experiment, the microblogging site has exploded onto the new media scene like a rotten egg dropped from a high window, and she's been gathering intelligence about it from her girlfriends. When I tell her that many celebrities' tweets are ghosted by staffers, she rolls her eyes. "That is sooo something you would say, Mum," she tells me. No arguments there. She tosses her head in irritation, as if my words were blowflies. But I can see from the look on her face that she's thinking about it.

Social media cost employers $2,700 a year per worker in lost productivity, according to one recent survey.[23] If they could put a price on parents' lost sleep, we'd have to declare national bankruptcy.

Parents angst about their kids' media use generally. But studies show our parental paranoia peaks around social media. That's understandable, given the interactive nature of the beast. We are all too aware of the risks—especially those posed by cyber-bullying and online predators. Yet most of the time we feel pretty powerless to do anything about it.

One recent study found that 71 percent of parents speak with their kids about online safety, but only half that many impose controls. (N.B.: Speaking of controls, the research was sponsored by an Internet filter manufacturer!) Lord knows, cyber-shit happens. In the case of online bullying, pretty much constantly, in fact. In Canada, the United States, and Great Britain, a third to one half of teens report being victims of targeted online abuse by peers. A study of four thousand children aged twelve to eighteen published by the Cyberbullying Research Center in 2010 found 20 percent admitted to having been repeatedly harassed, mistreated, or ridiculed by another person online or while using cell phones. "Mean or hurtful comments" and rumor mongering were the most common forms of abuse.[24] Australia purportedly has a much lower incidence. A government-commissioned report issued in September 2009 found fewer than 10 percent of kids aged ten to fourteen had been bullied online or via a cell phone, ranging upward to 20 percent of sixteen- to seventeen-year-olds.

Not surprisingly, victims of cyber-stalking are rarer still. Regarding other forms of inappropriate use, Australian experts estimate that 84 percent of boys and 60 percent of girls have been "accidentally exposed" to pornography online—curious that the boys have so many more "accidents"!—while 38 percent and 2 percent, respectively, have been deliberately exposed.[25]

Without underplaying these risks, the truth is that out-and-out abuse is probably the least of our problems as parents of Digital Natives. It's a bit like our overblown stranger-danger fears, when statistics show very clearly that family friends and relatives pose by far the greatest risk for sexual, emotional, and physical abuse of our children. Or, for that matter, like our fear of flying versus our lack of fear of driving home from a party. We tend to badly misperceive where the real dangers lie: in the mundane and familiar environments that surround us. Their very familiarity means we look through, not at, them. And therein lies the risk.

Consider for a moment the visual element of social media, which, as the name "Facebook" suggests, is pretty much the point of the exercise, especially as far as teens are concerned. Five hundred, seven hundred, even a thousand photos on an individual account are nothing unusual. On Flickr, aka "the World's Photo Album," more than thirty-five million users have uploaded more than three billion digital snaps, and more are added at the rate of three million every day.[26] One can't help but wonder: If a picture is worth a thousand words—how do we begin to do an audit on three billion?

For my money, what's really scary about all those social media

photo ops isn't the remote possibility that some pedophiles or predators may be stalking our children. It's the absolute certainty that our
kids will stalk one another—for hours and hours and hours on end,
through an endless labyrinth of fake tan, rippling abs, and plumped-
up lips. It is all innocent enough. Narcissism, as we know, is what
teens *do*. But that's exactly the point. It's the intersection between
what comes naturally (obsessing over image) and what the technology does best (producing and broadcasting those images to the world)
that makes it risky business. Personally, I am more afraid of a Year
8 girl who Photoshops her digital snaps to create "flawless features"
than I am of almost anything.

As the great moral philosopher Pogo the Possum once remarked,
"We have seen the enemy. And he is us." Then again, in a world where
44 percent of Internet users have an online identity different from their
identity in real life, that's arguably more complicated than it sounds.

At the University of Maryland, student athletes were sick of getting busted by their coaches, who have learned to stalk Facebook
for incriminating photographic evidence of pre-game carousing. By
September 2008, the situation had gotten so bad that students were
actually trying to ban camera phones from their own events. Zeynep
Tufekci, who teaches sociology to those students, is convinced that
social-networking media are making us more, not less, accountable
for our actions. "We're going back to a more normal place, histori-
cally," she observes—a place not unlike a small town, where everybody knows your business, whether you want them to or not. Identity
theft is no longer the issue, Tufekci argues—but preserving anonym-
ity may well be. "You know that old cartoon? On the Internet, nobody
knows you're a dog? On the Internet today, everybody knows you're a
dog. If you don't want people to know you're a dog, you'd better stay
away from a keyboard."[27]

Other observers worry that our meaningful relationships are being nudged aside by one-sided "parasocial" connections, such as Sussy's relationship with Taylor Swift or Zooey Deschanel: "peripheral people in our network whose intimate details we follow closely online, even while they . . . are basically unaware we exist," in the words of Danah Boyd, a fellow at Harvard's Berman Center for Internet and Society.[28] Social media have enabled an explosion of what anthropologists call "weak ties." But wither the strong ones? The deep ones?

And speaking of getting real, Flickr cofounder Caterina Fake—and no, I am not making that up—admitted recently that the ease of online sharing has made her slack about getting together with friends the old-fashioned way, in high-resolution reality. "These technologies allow you to be much more broadly friendly, but you just spread yourself much more thinly over many more people," she explained.[29]

And who wants to raise a stack of pancake people? My worst fear as a parent was that my kids might lose an alternative frame of reference—that growing up as Digital Natives, they would swallow the pancake paradigm whole and forget there were more nourishing ways for friends and family to connect.

The evening Sussy and I hunkered down in front of the fire with the boxes of family photos ("Whoa. Look at all those hard copies!" she cried) for a veritable festival of face-to-Facebooking was a good case in point. We devoured thousands of images, laughing, hooting, or blinking in wonderment just as we would have done online. But sitting side by side, passing pictures from one set of hands to another, created a different energy. We didn't simply consume the images, or allow them to consume us. Rather, they became catapults, triggers for stories and recollections, for the exchange of family and cultural history far greater than the sum of the individual parts. "Yes, darling, Grammy *was* a hottie back in sixty-nine," I agreed, my eyes bright with unshed tears. "No, I'm pretty sure that *was* her real hair."

The impromptu glee club I encountered that summer night around the piano evoked similar longings: more than a nostalgia for the real, it was a déjà vu about the real, I reflected, as the playlist skidded freakily from *The Jungle Book* to Death Cab for Cutie and back again. "I had no idea Mason Reeves could play the piano!" I exclaimed to Anni after the group dispersed that night. "To be honest, neither did I," she admitted.

"Was it okay? I mean, you all looked like you were having fun . . ." I trailed off.

"Fun?" she spat back. "You must be joking! It was awesome."

As we've already observed, Bill's exile from MSN, Facebook, and his anime stash propelled him out of the door faster than a bullet from one of his beloved first-person shooter games. My dread was that he would simply make a beeline for The Beast, hunkered down at Vinny's only a few streets away. And he did too—at first. Within a week or two, his separation anxiety seemed to dissipate. He started spending more time at the beach and pool, catching up with friends he hadn't connected with since primary school. Matt, for instance, who was now a serious trumpet player, and Tom, the older brother of Bill's gaming buddy Pat, who had recently taken up jazz piano. They were both studying with the same teacher, a saxophonist named Paul Andrews, Bill reported. And so began the prelude to his renewed interest in the saxophone. Any chance that he could start lessons again? he asked me soon afterward.

I pretended to consider it—no sense ruining everything by showing my approval—and agreed to a "trial lesson." I came in at the end of it, just in time to see Andrews nod his head curtly.

"So, tell me. What do you want to be?"

"A musician," Bill replied without hesitation.

("WTF?" I was screaming internally.)

"Uh huh." Andrews nodded again. "Well, practice, focus, listen, learn . . . and you can be."

Up to that point, Bill had barely picked up his instrument in two years. From that point, he has hardly put it down.

In the ensuing weeks and months after that pivotal first lesson, I watched my son evolve like a human Pokémon from a surly, back-talking gamer to a surly back-talking musician in-the-making. (LOL.) To this day, Bill insists that it wasn't The Experiment that changed him. It was the friends, and the teacher they'd led him to. "Ah. I see," I reply.

"The technology ban was nothing but a trigger," he adds, a little less certainly.

"Ah, a trigger," I echo. (Bang, bang! I think to myself. Got 'im!)

Sussy ended up switching friendship groups too. But she was under no illusions about the role The Experiment played in bringing about the change—nor was the Year 10 coordinator at her school. "Sussy has shown a marked improvement in terms of not only her demeanor but also her organization," she wrote to me in April 2009. "She seems to be adapting to having to complete any electronic work at either school or at her friends' house and her uniform has been consistently pleasing." (A big clean-out of her bedroom uncovered not one, not two, but three "lost" name tags, which she took to wearing in a row on her blazer pocket, like war medals.) "Sussy seems to be a happier student who is becoming more independent and taking more responsibility for her learning."

Loss of Facebook (not to mention loss of MSN and MySpace) seemed to increase her focus generally; at the same time, it put her out of the loop with her old friends. "With Jen and Cat and that kind of group, you figure stuff out on the computer, like sleepovers and stuff," she explained to me at our midterm interview. These invitations happened spontaneously, usually on the spur of the moment, in fact, with little or no notice. If you blinked—or, more to the point, if you went offline—you missed them. The girls in Sussy's new group at school didn't operate like that. "We planned a sleepover a week in advance!" she told me proudly, and slightly incredulously.

Sussy's Experimental coping mechanisms differed from Anni's and Bill's significantly. The older kids took the opportunity to go out more—shopping, visiting, or clubbing in Anni's case, and hanging out at the pool or jamming in somebody's garage in Bill's. Sussy had fewer friends who lived in the neighborhood, so she faced major transportation issues. Her best girlfriend, my goddaughter Maddi, lived in Melbourne. Her closest boy chum, Andy, had just moved with his family to England.

Partly for these reasons, her overall media time budget probably remained unchanged.

She clung to the landline like a drowning teenager to a life raft. After school, she'd install herself in the family room, echoey and airplane hangar–like now that it had been clear-felled of its media and their bulky accoutrements, and hold court before an unseen audience for two or three hours at a clip. She assured me that both Maddi and Andy had their parents' permission to ring her as often as they liked; it seems they had magic Internet landlines that made long-distance calls for free "or just about."

"What if you need to ring them?" I wanted to know.

"Easy. I just send them a signal—I ring once or twice and then hang up. Really, Mum, we've got it all figured out."

Many people have asked me if there was ever a moment during The Experiment when I was tempted to quit. Not counting April 25, the day I received a phone bill for $1,123.26, I can honestly say, no. Not at all.

Digital Immigrants use technology to achieve specific ends. Digital Natives breathe technology in order to . . . well, breathe. To exist. Before The Experiment, Sussy had pretty much lived online. Now she was pretty much living on the phone. Cleverly, she also used it to gain access to banned media. "Google 'Nick Jonas!'" she'd bark into the phone to Maddi, when the need to know the details of Miley Cyrus's relationship status grew unbearably urgent, or, "Check my Facebook!" (the girls regularly, and companionably, hacked each other's accounts anyhow), or, "Message Andy and tell him to ring me at eight my time." Maddi was now more than a best friend. She was Sussy's personal remote outsourcer, carrying out her digital bidding with terrifying dispatch.

Their relationship changed in less obvious ways, too, during those marathon conversations, and so did her connection with Andy. "On MSN, you're kind of almost waving at people. You get introduced, and it's like hi and LOL and ILY and stuff . . . but you never really get to *know* them," she explained to me. "On the phone, it's totally different. It's like D&M.* You get close. You get tight."

Anni agreed with that. "I think it's definitely a more intimate thing. Talking on the phone isn't the same as face-to-face, but it's a step up because you have the tone of voice and everything, and you can infer so much more than from stuff that's typed up. Texts are like decoding messages—so hard to interpret! It's always like, what did they mean by that? Are they kidding? Are they being nice? Are they being condescending?"

Do you think people are more honest in phone conversations? I ask Suss. "Yeah, and you explain stuff. And they ask questions and stuff, and . . . I don't know. How am I supposed to know? CEEBS!" (Munted family acronym deriving from CBF—couldn't be f#*#ed.)

"Only connect."

 "Only connect."

 "Only connect."

 With no new message alerts to distract me, I keep going back to Forster's mantra . . . or was it a plea?

What lies behind our mania for media, anyhow? Before, I assumed

---

* Deep and meaningful.

it was something to do with our insatiable appetite for entertainment, for data, for distraction. Now that we were in the purge stage of the bulimic cycle—digitally, that is—I wasn't so sure. Maybe our most implacable desire, our most deeply human yearning, is simply to achieve contact. To . . . connect.

James Harkin, author of *Lost in Cyburbia*, observes correctly that our new media deliver "connection" in an entirely unprecedented way. Old media delivered stories. Our new media—e-mail, IM, social networking, microblogging—deliver *people*. Social intercourse. Contact. The technological equivalent of holding hands, or even making eye contact.

This drug we're craving . . . could it simply be each other?

For young people, the evidence is clear that Facebook is literally better than sex. And I do mean literally. According to Internet tracker Hitwise, visits to porn sites dropped by a third between 2005 and 2007. By 2009, the industry outlook was so grim that *Hustler* publisher Larry Flynt got on his knees to the federal government seeking a $5 billion "stimulus package." (Ew.) Evidently the most precipitous loss of libido has been among users aged eighteen to twenty-four—the same demographic who, not at all coincidentally, have invented the joy of friending.

Not only are social media better than sex, they reproduce more efficiently too. Network theory reminds us that the number of possible connections between points on a network rises much faster than the number of points themselves. (Two people with fax machines can only talk to each other. But five people with fax machines create twenty possible channels; and twenty people with twenty fax machines, 380.) The resulting explosion of connectivity, Harkin tells us, makes networks exponentially powerful. Just like kids, really. You give them an inch, and they take a gigabyte. When advertisers exploit this principle, they call it "going viral." When Facebook or LinkedIn users do the same, we call it social or professional networking. When our kids do it on Facebook Chat or AOL or Google Talk, we call it a huge waste of time. Hmm.

Seriously, it is exactly this network effect that turns a teenage birthday party for twenty close friends into a drunken, gate-crashing horde. Or that makes "sexting"—the mass forwarding of sexually explicit photos via cell phones—such an insidious and effective form of cyberbullying. The network effect enables collaborative enterprises such as Wikipedia and YouTube (the latter so huge it accounts for 10 percent of the Internet's entire bandwidth) and propels the careers of out-of-the-box upstarts such as *Britain's Got Talent*'s Susan Boyle, or, for that matter, Barack Obama (an unabashed BlackBerry addict, let us not forget).

James Harkin tells the story of Shoreditch Digital Bridge, a project that provided free Internet access to people living in a public housing estate in east London. As an afterthought, the project managers also decided to offer residents access to CCTV surveillance images. These security cameras were already providing twenty-four-hour monitoring of communal areas, so why not broadcast it as a channel like any other? Nine months later, a leaked report showed that residents watched the closed-circuit security channel as much as they did prime-time broadcast television. Literally more people tuned in to view each other than to see *Big Brother*.

The medium has become the messenger. Caught in the sticky tendrils of Web 2.0, we stand transfixed, not by "data" or even "entertainment," but by ourselves. What bedazzles us most of all is not the shock of the new but the shock of recognition. Of affirmation.

"Only connect," indeed. Could Forster have imagined how relevant that commandment would become in the age of Apple? For that matter, could he have imagined a $1,123.26 phone bill?

Keeping oneself in the loop is all well and good—but where will it end? Well, that's the thing about a loop. It *doesn't* end. Futurist Raymond Kurzweil predicts a coming techno-evolutionary quantum leap that he calls "the singularity." Beings of this new age, half human and half machine, will be endowed with enhanced, yet alien, brains and near-immortal lifespans.[30] Gosh. Perhaps we will call them "teenagers."

May 3, 2009

Have stocked up on CDs and tapes—tapes!—from the library, and Bill loving his new turntable. ("What do you like best about it?" "The crackle!") Also confesses he enjoys the turning part of the turntable— just watching records revolve. A novelty, I guess, in a world of CDs and audio files. Has raided various friends' parents' closets for LPs and amassed huge collection already: Doors, Bob Marley, Stones, Beatles.

I sneak down to his bedroom sometimes just to see him lying on his bed, reading and listening to seventies music—"Like something out of *Back to the Future*," Sussy hisses.

Girls and self magged out with *Wish*, *The New Yorker* (great Lily Allen profile—now there's a girl who knows about oversharing . . .), *Sunday Times Magazine*—local gossip column a total crack-up—and *Girlfriend* ("Is Your Boyf Too Metro?"). *That's* entertainment.

May 4

Listening to music and doing nothing else at the same time? How weird is that?

But am doing it anyway: practicing Thoreauesque attentiveness listening to my new Leonard Cohen CD.

Overheard at dinner:
BILL: I find school an intrusion on my saxophone practice.
SUSS: I find your saxophone practice an intrusion on my life.

Overheard after dinner:
SUSS (ON PHONE—WHERE ELSE?): Oooh, sorry! Well, break a leg! ILY!!
ANNI: What's up? Is Maddi out on a date or something?
SUSS: Nah, she's just making her entrance in the school play.
ANNI: WITH HER CELL PHONE SWITCHED ON??
SUSS (WITH DIGNITY): Of course not. It's on vibrate.

May 9

The Fremantle public library rocks. Hadn't been there since I used to take the kids for story time and a hand-stamp back in the nineties. Go on Saturday afternoons now, after chores, and today check out six items JUST FOR ME! Weird Norman Mailer book about Jesus, *On Kindness*, some yoga book for the decrepit (i.e., me), *The Namesake*, *The Death of the Grown-up* (seemed appropriate under the circumstances), and a Thelonious Monk CD. Such a thrill of anticipation, bustling around making tea and laying the fire like some jazz-loving Jane Austen character.

May 11

Suss home from school today excited about Emily Dickinson poem she read in English. Amazed and slightly spooked when I produced *The Complete Poems* and showed her the concordance at the back. "It's sort of like CtrlF for fossils," I explain. Still couldn't find the poem she wanted. "You could always ask Maddi to Google it," I suggest evilly. "NO!" she cries (though am still not sure why).

Frances rang about coming over to work on captions for her book. Remind her I don't have a laptop. "No worries," she says. "I'll bring mine!" Explain, gently, that she can't do that either. "But it's not as if we can't do them in longhand."

"Tell me you're joking," she pleads.

Kids hoot with laughter when I tell them.

Seeing grown-ups get their comeuppance: Does it get any better than that?

May 13

Bill's b/day tomorrow: SWEET SIXTEEN!!!!

Gifts purchased: Coltrane two-disc set; Coltrane songbook; ten Murakami books (including one collection called *Birthday Stories*), and two hipsterish T-shirts selected by A. & S., who are getting him LPs from op shop.

Last year he got a Nintendo DS.

Yelled at A. for playing Snake in the car on her phone. "I'm not in the house!" she snapped. But put it away.

May 14

"Did you ever think you'd be excited about getting eleven books for your birthday?" I ask Bill happily. I think we all know the answer to that one.

Lovely Japanese b/day dinner at Fuji followed by bubble tea at Utopia (an Asian fast-food joint and B.'s self-proclaimed "spiritual home"). He ordered mocha bubble tea with large chocolate sago, and Suss had an Oreo smoothie. Weirdness! A. & I amused ourselves examining the plastic sample meals. Laughed nonstop and had the best time just hanging out together.

Couldn't help but compare it with A.'s eighteenth b/day dinner last October, when B. & S. begged to be allowed to sit in the car—i.e., access iPod and DS—the moment meals were cleared.

May 16

Much of day spent munching on a devo burger, as Suss would say. ["Devo" = devastated.] Can't be PMS as no longer have the M. May have gone overboard on the renunciation thing. Realized today on return from early-morning walk—new routine—that, in addition to giving up media, I have renounced my cigarette (had been allowing myself one a day, like a vitamin) plus—thanks to Atkins—alcohol, sweets, bread, pasta, and rice. Sheesh. As fate would have it, am also celibate, so there is literally nothing left to give up.

S. holed up in A.'s room making her first-ever mix tape. We are going to play it in the car. S. also spent FOUR HOURS deep cleaning—like flossing, practically—her bedroom. Was almost frightened to see her in action. Later showed me several neat pages of math revision for tomorrow's test. WTF?!

May 17

Bill to me in car: "So who's this Kafka, anyway?"

Death of Atkins celebrations tonight with Mary and Grant and family: spaghetti carbonara, sausage risotto, cookies, and Suss's amazing chocolate cupcakes.

After dinner, the assembled kids (aged sixteen, fifteen, and fourteen) played hide-and-seek. B. folded himself up like an unassembled IKEA wall unit and hid on a closet shelf.

May 20

S., enigmatically: "I think The Experiment is finally beginning to work on me." Then the phone rang, alas, and she fled.

When questioned later, added: "I'm doing shiz now—baking cupcakes, writing in my journal [she has a *journal*?????], planning a novel [she's planning a *NOVEL*?????]. You know, like . . . shiz."

Eat, Play, Sleep

It is hard to provide and cook so simple and clean a diet as will not offend the imagination; but this, I think, is to be fed when we feed the body; they should both sit down at the same table.
—WALDEN, *chapter 11*

y first child, Anni, was a breast-fed, demand-fed baby. To varying degrees, all my children were. Breast milk was never really fast enough on the draw for Bill, although he humored me for six or seven months (if fussing and spitting up and looking continually aggrieved is your idea of humoring). Suss wasn't exactly a poster child for La Leche League either. She'd bonded early to her ba-ba and, given the two other toddlers I was wrangling at the time, that suited me fine. But all three were the beneficiaries of the demand-feeding orthodoxy that had ruled supreme in the parenting literature ever since Dr. Spock.

Demand feeding—basically, fulfilling the nutritive needs of the child as they happen over the course of the day (and far into the night, as the case may be)—makes such intuitive sense that many new parents never think to question it. I certainly didn't. Putting a baby on a schedule was something they did in the bad old days of the 1950s. It went along with smoking in pregnancy, or "airing" the baby in all

weathers, as if it were a feather quilt. Four-hourly feeds were rigid
and ridiculous, I would have told you at the time. A baby was a living,
changing organism—not some kind of automaton. (Never mind that
I'd been raised that way myself without obvious mechanical damage.)

I'm not saying that demand feeding isn't a good practice. What I
am saying is that no one ever told you when it should stop.

Anni was about twelve when I asked her, as I did most nights, if
she was hungry for dinner yet. "Not really," she admitted. "In fact,
I've *never* actually been hungry."

Eighteen years after commencing lactation, I was forced to admit
that my kids were still being demand-fed—though they called it
"grazing" now that they'd gotten as big as horned ruminants. It was a
pattern I'd been aware of for some time. But The Experiment revealed
something about it that I'd never considered before: It wasn't only my
direct failure to take a hard line on four-hourly feeds (aka breakfast,
lunch, and dinner) that was to blame. It was also the way we'd structured our media ecology . . . or, more accurately, the way we'd failed
to structure it.

As sunflowers turn to the light, our family's primary attention
had turned toward our screens over recent years. Over roughly the
same period of time, the social function of family meals had pretty
much crumbled. Coincidence? The more I thought about it, the
more I thought not. It's true, we had always been a demand-fed family. But once upon a time we had also been a family that sat down
to meals together, around a table, in a technology-free space—aka
the kitchen—that was the functional, beating heart of our home. In
recent years, we'd undergone a bypass, big-time. People had started
to grab a bite where they could: at their, quote-unquote, workstations
(where 95 percent of their time was spent slacking), or on the way to
school, or training, or social events.

My friend Susan, who grew up on a sheep station in New South

Wales, used to wonder why people didn't just eat pellets, like livestock. Watching my kids graze, I often wondered the same thing myself. We didn't "dine" anymore: We fueled.

I see now that I was part of the problem—a really big part, it pains me to admit. Sure, their lives had gotten busier (or so they kept telling me). But mine had too (or so I kept telling myself). Sitting down to meals was a luxury that working families like ours couldn't afford. Everybody knew that. There simply wasn't time to squander on sit-down home-cooked meals, like some sepia-toned tableau from the golden age of grade-B television. I wasn't entirely pleased to see the collection of cereal bowls and teacups littering the backseat of our car most mornings—or to hear the crash of crockery and the clatter of coffee spoons each time we went over a speed hump—but I accepted it as part of the deal of having teenagers.

As primary-school kids, they'd pulled *me* out of bed in the morning. Now I find that extraordinary. (One day in the not-too-distant future, perhaps when they are changing my diaper, I may see things differently.) Back then, cooking breakfast—eggs and toast, oatmeal, French toast, and fruit—and sitting down together to eat it was the easy part of the day. Now it seemed to take every ounce of my maternal energy just to prize them from their lairs. I'd become a human snooze button, going off at regular intervals. By the time I simply started yelling—about five minutes before blast-off—a smash-and-grab approach to the Most Important Meal of the Day was the best anybody could hope for. I'd facilitate, grouchily—thrusting hastily made peanut-butter sandwiches into people's backpacks as they lurched past, or sloshing three travel thermoses of tea in the general direction of the car.

At dinnertime, I still tried hard to bring all the stakeholders to the table. Sometimes, against the odds, I succeeded. But between Bill's water-polo schedule, Anni's social schedule, and everybody's repeated pleadings for "chill-out time"—a thinly veiled euphemism for sitting, trancelike, before their respective screens—achieving a quorum had become a rarity. And even when I succeeded in corralling them to the table, my best diplomacy failed to keep them there. Most nights, Bill would be washing down the last bite of his dinner with a third glass of milk by the time I picked up my fork. "Thanks, Mum!" he'd call cheerfully as he dumped his plate in the sink and turned fervently once more toward Mecca (as I'd begun to think of The Beast). He'd sit back down if I insisted, of course. But forcing a boy to be sociable is a bit like teaching a pug to break-dance. The quality of result simply isn't worth the bother.

"So, how is school?" I'd ask.

"Good," he'd say.

"What are you reading in English at the moment?" I'd ask.

"Books," he'd say.

"Any book in particular?" I'd ask.

"Not really," he'd say. "Can I go now?"

The girls didn't eat quite as fast, but they, too, tended to react roboti-cally to my crude attempts at conversation. It was almost as if we'd corporately become one of those joyless married couples you see at res-taurants, eating purposefully and in total silence. Together in flesh but on entirely separate hemispheres of the spirit. "Other families eat din-ner in front of the TV," Sussy would sometimes remind me wistfully.

They weren't the only ones without an appetite for all this. Per-sonally, I'd have preferred to read my novel most nights too, though admitting that even to myself made me feel guilty and wrong. You know, normal. I'd banned the girls from texting at the table—indeed, had never allowed it—but most nights I could feel their attention being tugged by their absent media, like a spider sensing a fly at the far edge of its web. They were, all three, so clearly tolerating our time together. It was as if our kitchen had become a transit lounge, an uncongenial place of temporary detention; a pit stop separating them from the virtual places they would much rather be.

The demand-feeding thing was the brittle icing on an already stale cake. And the evidence lay strewn near every monitor: wrappers, crusts, crumpled cans and cups, fruit peel, the petrifying remains of instant noodles, cookie crumbs, half-empty (never half-full) water bottles, and the occasional Pollockesque glob of gum. The low-level gratification of binge snacking was clearly the perfect accompaniment to the low-level gratification of binge connectivity. Except that "binge" wasn't really the right metaphor, implying as it did a time of abstinence or purging. More accurately, the children were on a sort of dual IV drip: data through one tube, Doritos through another.

Everybody knows that food fuels activity. But as new mothers learn the hard way, the connection between food and rest is equally direct and every bit as dynamic. A hungry baby will not sleep, a sleepy baby will not feed. A baby who sips through the day and catnaps through the night may be healthy enough, but is rarely robustly happy. It's not just his moods that are short. His attention span is too. This is why a child whose feeding–waking cycle is messed about, or never given the chance to establish itself in the first place, will be a child who struggles to settle down to bigger things, whether playing peek-a-boo or learning to parallel park.

The connection between eating ravenously, sleeping deeply, and concentrating in a focused and sustained way—whether on work or at play—is so obvious when our kids are babies. We spend half our lives as parents shoring up those boundaries, keeping to the routines that we know will help our children to thrive. We are happy to let our own lives descend into blurriness for a while, missing sleep and meals and sex and the sustaining goodness of uninterrupted work, to ensure that *their* lives have clipped edges and crisp demarcations. But it's not just babies who do best under conditions of clarity. We all do. Teenagers, arguably, more than most.

Watching my kids get swallowed up into the maw of multitasking, I observed the onset of a condition I came to think of as "blobbiness." It was *not* next to godliness.

I'd made quite a study of blobbiness over the past two to three years, and it seemed to me its symptomatology—personal untidiness, poor eating habits, disordered sleep–wake patterns, ineffective time management, ineffective "stuff" management (losing personal items, forgetting to take lunch, losing track of money) and mood changes positively free-associative in their movements to and fro— was all about boundaries or, more precisely, about not having any. If I were a physicist, I'd use the term "entropy" instead of blobbiness: the tendency of a system to move toward randomness, loss of heat, and decreasing differentiation of parts.

The parents of teenagers tend to employ another word for all this: normal. This is what adolescence is all about, we tell ourselves and each other. The popular media support us in this view, reminding us constantly that teen brains are different, that we should not expect "adult" (read: responsible) behaviors from them before the age of twenty-eight or even thirty-two, that teenagers have "always" been messy, lazy, clueless layabouts.

In fact, most of the symptoms of adolescence—and we do treat it as a disorder—are nowhere to be found in many other cultures or in earlier historical periods. The tendency of young people to fall in love or lust and to behave from time to time with a certain impetuosity (i.e., like bloody idiots)—these are universals, it seems. But the rest of it is acquired behavior, the consequence of greater leisure, greater affluence, a prolonged period of schooling—and our correspondingly lowered expectations that our teens will contribute productively to family or community. The more childish and irresponsible behavior we accept from them—because that's the way they "are"—the more childish and irresponsible they have become. Truly, they are only following orders.

And it is important to grasp that those orders are coming from our culture as a whole. They are what sociologists call a "social script" or a "social construction." Your, quote-unquote, parenting philosophy, or mine, is not primarily a work of your imagination or will or wisdom, although hopefully all these make an appearance at some point. It's not like a craft project that you design and complete in your spare time, although—again—we are encouraged to think of parenting this way. In ways that are terribly important, we actually do get handed a set of Operating Instructions (albeit encoded at a very high level of abstraction) at our kids' births. The instructions don't tell us what to do at every step along the way. But they do tell us what the basic game plan is, and what object we are aiming for. They key our expectations for achieving—or enduring—what we have learned to call "developmental milestones."

The climate change we have witnessed in the global media ecology over the past fifteen years hasn't created the blobbiness we observe in our kids. But it has certainly intensified it, and in some ways legitimized it, pushing our children to the next blobbiness level. The explosion in connectivity (if not always in communication) that has enabled the 24/7 lifestyle to which our Digital Natives incline has detonated a whole new set of land mines, reducing an already besieged set of personal and social boundaries to tatters. Blobbiness is no longer a stage—a difficult phase to be tolerated until it "passes through," like a storm, or a kidney stone. In the present media ecology, blobbiness is not exceptional in the least. On the contrary, it rules.

Anni's appetite issues were blobbiness personified. Snacking on soy crisps was a way of life for her. She never felt entirely hungry, nor entirely satisfied. No meals were fully eaten, and no fasts were fully broken—only sort of dented. Snacks were not the station between meals, but the destination itself, nibbled by the handful here and there throughout the day, as though she were a house cat. Anni's version of dry food was not uniformly unhealthy. She did have a tendency to hoe into chips and crackers and sweets—indeed, her utter lack of impulse control around junk food was more doglike than feline—but she also ate tons of yogurt, cheese, fruit, and salads. Overall her diet was not unbalanced, it was just . . . well, blobby.

Nor was there any discrimination in where she consumed her food. Any room in the house would do: the lounge, the TV room, her bedroom, of course (it often bore the look and smell of a recently vacated school cafeteria), and even the bathroom, where it was a common sight to encounter a half-consumed piece of Vegemite toast cheek by jowl with a hairbrush, or an empty juice glass next to a bottle of Sussy's "Age-Defying Foundation" (at fourteen, I guess you need all the defiance you can get).

Further blobbiness was evident in the way food was matched to time of day, with Rice Bubbles slurped up at dinnertime and pizza cheerfully consumed cold and congealed on the way to school. A few months back, Bill had attempted to solve this problem by buying a bulk carton of mee goreng instant noodles from our local Asian grocer. This way, he reasoned, his favorite snack could be promoted from the fifth basic food group to the *only* basic food group. "Man cannot live by mee goreng alone," I advised sternly. (Apart from the utter lack of nutrition, the salt content made Vegemite look like blood-pressure medication.)

"What are you talking about?" he replied. "You have to add boiling water."

I'd wonder sometimes if the whole thing really *was* my problem, as the kids insisted. "Grazing is good for you! All the experts say so," Anni assured me between chomps of her Chicken Crimpy snack crackers. "In prehistoric times, that's the way all human groups got their sustenance." What? When Crimpy Chickens still roamed the earth? I was pretty sure the environment-of-evolutionary-adaptedness argument had little relevance to a food chain dominated by blue sports drinks and flavor [*sic*] packets. But it was also true that I'd grown up in a culinary Stone Age, where you sat down with your family for dinner every night and ate what was put before you, whether animal, vegetable, or (in the case of my mother's dreaded Iron Man Casserole) mineral. Processed foods were few, and the whole idea of individually wrapped snacks was still a rough beast waiting to be born.

My kids cannot imagine a world without muesli bars, or cheese strings, or yogurt you squirt from a tube, or fruit you can roll into a single, cigarette-size cavity ripper. In my day, your mother was as likely to put a fun-size Mars bar in your lunchbox as she was to serve canned hash to the bridge club. We really did eat fruit for a snack, I used to try to explain to the kids when they were little—not because we loved fruit so much but because there wasn't an alternative. "Poor Mummy! Couldn't anybody flatten it for you?" they'd ask plaintively.

Sure, if we'd had Dunkaroos the world might have turned out to be a different place. But the point is, we didn't have Dunkaroos. We just had to soldier on anyhow.

The rules we observed for meals were to blobbiness as basic training is to a Montessori school. This was especially true for the evening meal, which, in white, middle-class Anglo households, predictably consisted of meat, a starch, and two vegetables. On the dinner plate, as in one's bedroom, there was a place for everything and everything in its place. (Only foreigners messed their food together in untidy piles.) As children, my sister and I even preferred to *eat* in a bound-aried way: first the meat, then the carrots, then the peas, saving what we liked least for last, the better for poking into a potato jacket, or scattering artfully beneath a chop bone.

Eating in front of the television was not unheard of, especially on Sunday night, when Ed Sullivan was on. But it was a vice my mother disapproved of. "Too much upsetment," she decreed (as if eating our tuna patties off a TV tray would cause us to run amok, ripping the plastic covers from the formal lounge with bared canines).

Was I nostalgic for all this? I was not. As a parent, I'd never wanted to bring back the halcyon days of casseroles held together with mush-room-soup mucilage and family mealtimes so rigid and ritualized you could practically hear your arteries harden. Lord knows, I had no appetite whatsoever for the whole meat-and-three-vegetables thing. It had been bad enough having to eat it. Having to produce it night after night would have really stuck in my craw. At the same time, I was determined to use The Experiment as an opportunity to combat the blobbiness epidemic that had overtaken our eating habits. I wanted to try to bring more structure to mealtimes, to put more energy into appreciating what we had on our plates. Or noticing it, even.

In a recent Australian study, four in ten mothers describe dinner as an "unpleasant experience," with the meal usually ending in an argument. At the same time, 76 percent agree that sit-down meals strengthen their family's communication (and possibly its vocal cords), according to a recent survey of more than 16,000 mothers nationwide.[1] Contradiction? Not necessarily. Maybe the experience of being together as a family is a bit like eating your spinach. As Popeye might have observed, that which doesn't kill us makes us stronger. Like it or not—and clearly four out of ten of us don't—family meals are consistently correlated with positive outcomes for children. And not just slightly positive outcomes. Ridiculously positive ones. Kids who eat family meals five to seven times a week get better grades, have a sunnier outlook on life, have significantly fewer problems with drugs, alcohol, or nicotine, and seem almost magically protected from developing eating disorders. They also—surprise!—have healthier diets. Recent research from the UK Department for Children, Schools and Family found a direct link between the frequency of family meals and high school leaving scores, while a study published in the *Journal of Adolescent Health* in 2008 uncovered a clear, inverse relationship between "eating together as a family" and risky sexual behavior. Weirdly enough, simply having supper together was as protective against unsafe sex as "doing something religious together."[2] Then again, maybe it's not that weird.

It's not the "postcode effect" either (where socioeconomic class is the underlying determiner of advantage). Researchers in study after study have controlled for demographics and the findings remain. Rich or poor, middle class or underclass, highly educated or barely educated, families that eat meals together are dishing up a smorgasbord of advantages for their kids.

These facts are hardly news—although the media love nothing better than to give parents a serve on the topic. Or mothers, more accurately. In most accounts, the demise of the family meal is attributed to the usual suspect: feminism—or, as it is more decorously described, "women's participation in the workforce" or "the dual-earner family." The implication is that when mothers work, families, like chickens, go free-range and slightly feral. Yet in Australia—where the full-time workforce participation of women with children is much lower than it is in the United States and the UK—a mere 11.42 percent of mothers report that their children usually eat at the family table. Remember, too, that we are talking about where and how family members eat, not about who (or what) is doing the cooking. The effect is exactly the same, whether it's a roast with all the trimmings, a stir-fry with fourteen intricately diced and unpronounceable vegetables, or burgers and fries eaten straight from the wrapper.

Instead of blaming mothers who work outside the home, maybe we should be looking more carefully at the media at work within it. Researchers from the Pew Internet & American Life Project found that families with "multiple communication devices" were less likely to eat dinner with other household members, and they also reported less satisfaction with their family and leisure time.[3]

The speed with which digital devices have invaded our domestic lives has left sociologists and other family researchers with virtual egg on their face, scrambling to keep pace with change. Reading research that is more than two to three years old is like traveling in a time capsule. (The dangers of "chat rooms?" "Online bulletin boards?" Who even knows what those terms mean anymore?) Even today, studies of the impact of technology on patterns of family food consumption focus almost exclusively on television—and this despite the fact that TV's market share is in steep decline among older children and teenagers. These findings are still worth examining—not only for what they tell us about TV per se, but for what they suggest may be true about screen time in a general sense.

Eating the evening meal in front of the TV, according to research conducted by the Nestlé Corporation in 2009, is almost twice as common as eating at a dining table.[4] The big question is, Does it matter?

Nutritionally, the answer is yes—although not by a huge margin. In a University of Minnesota study of five thousand middle and high school students, researchers found that teenage girls who ate alone typically consumed fewer fruits, vegetables, and calcium-rich foods, and more soft drinks and snacks, than girls who ate with their parents. They also took in 14 percent more calories. (As I've said, plenty of other research confirms that family meals protect girls against eating disorders.) The effects were similar, though less striking, for boys. Yet researchers noted that, compared with not eating family meals at all, eating meals together in front of the television was definitely associated with better eating; kids of both genders showed "high intakes of total vegetables [and] calcium-rich food, and greater caloric intakes."[5] Did socioeconomic class have anything to do with it? You bet it did. As might be predicted, more affluent families were less likely to report TV viewing during meals. But the general patterns held even when demographics were taken into account.

Overall, researchers concluded that "watching television during family meals was associated with poorer dietary quality among adolescents. Health-care providers should work with families and adolescents to promote family meals, emphasizing turning the TV off at meals." A study of families with preschool children, titled "Positive Effects of Family Dinner Are Undone by Television Viewing," found . . . well, I guess it's pretty obvious what it found.[6]

What about the psychological payoffs of the family meal? Does TV reduce the resolution here as well? A study published in the journal *Young Consumers* in 2008 argued that parents practically have a duty to capitulate to kids' demands for TV-enhanced meals. Mothers and fathers who refuse to do so, the authors argued, risked creating "social distance" within their families. "By joining her children at the television," a mother has the opportunity to engage with them while developing "a common interest," they noted, adding that "this communication can be seen as a way of maintaining love and relatedness in the family."[7] I hope I am not being cynical when I observe that *Young Consumers* is a journal devoted to "responsibly marketing to children."

On the other hand, if television helps bring teenagers to the table, it may be worth a look. Even researchers at the University of Minnesota conceded that "adolescents unhappy with family relationships"—i.e., the kids who arguably need parental contact the most—"may be more likely to participate in family meals if the TV is on and conversation isn't the main focus." One subject, seventeen-year-old Christina, complained that a media-free dining experience was just too boring. "It's fine at the beginning when Dad asks what we've done at school" but it quickly "gets boring without any music on or anything. If you eat in front of the telly, you have something to occupy your mind."[8]

Yet if conversation *isn't* the magic ingredient that gives the family meal its transforming power, it's hard to know what is. If eating dinner in silence in front of *Wheel of Fortune* qualifies as a "family meal," what about all those breakfasts we used to bolt in the car on the way to school? Did they count too? After all, we were all gathered in the one spot. It just happened to be moving at thirty-five mph. Some experts have suggested that the real secret to the family meal is simply that it gives parents a daily opportunity to "visually assess" kids for potential problems. Others concede that its power, while undeniable, remains mysterious—possibly even unknowable.

In 2008 pediatrician Katherine E. Murray found that family meals and family nutrition both declined significantly in households where teenagers had a television in their bedroom—and almost two-thirds of her sample did. They also engaged in less physical exercise, consumed more soft drinks and fast food, and read and studied less. Girls with bedroom TVs, public-health researcher Daheia Barr-Anderson found, spent almost an hour less a week in "vigorous activity"— extreme channel-surfing excepted—and ate an average of three family meals a week or fewer, compared with just under four meals for other girls.[9] For male teens, physical activity wasn't affected but school performance was. Grades for boys with TVs in their bedrooms were 10 percent lower than peers without TV. And boys, interestingly, are more likely to have their own televisions in the first place.

Among the multitude of things the family-meals literature *doesn't* tell us is whether the benefits increase arithmetically with time—if twenty minutes around the dinner table is beneficial, are forty minutes verging on miraculous?—but heading into The Experiment, it seemed safe to assume that more of a good thing was probably going to be . . . well, a good thing. Because we had always been a family that ate meals together, and did so without the benevolent assistance of television, I was looking to The Experiment as a way of extending the experience in both quantity (time spent) and quality.

Admittedly, we were coming off a pretty low base. I would definitely have put up my hand along with the 40 percent of Australian mothers who find mealtimes unpleasant AND the 67 percent who believe they are good for us anyhow. Most nights, I'd put a fair amount of effort into preparing a meal. Nothing lavish—like most teenagers, mine are allergic to lavish—but in the main nutritious, balanced, and quasi-palatable. When they were little, I scurried around making special child-friendly dinners. In fact, our nightly fare was not unlike the Kidz Menu at a down-market family restaurant: i.e., heavy on the chicken nuggets and carrot sticks, light on the line-caught trout and mushroom rillettes. These days, the experts tell you this is exactly what you shouldn't do. Children should be offered adult food from the git-go, and if they don't like it, let 'em eat multigrain bread.

But I have to say, though my kids' palates were definitely stunted, not having to engage in force feeding meant that most of the time I enjoyed our meals together. Still, as the children got older, I did begin to worry. Would they reach adulthood squirting ketchup like crack-crazed graffiti artists and removing the "crust" from their fish sticks on grounds of "spiciness"? Somehow or other, they eventually moved on. Today they are able to enjoy most foods, with one or two limitations. Anni doesn't like meat. Bill doesn't like vegetables. And Sussy isn't really into cutting things up, or for that matter chewing. But hey. You can't have your Thai fishcake and eat it, too.

Going into The Experiment, my main concerns about our family mealtimes were: first, low appetites on account of the ridiculous amount of after-school snacking (sorry, "grazing") going on, most of it in front of a screen; and, second, speed-eating. The latter was a term I learned much later, in doing the research for this book. It was a relief to find there was an actual word to describe the practice that had been poisoning the ambience at our family meals, like an overboiled head of cabbage, for years.

"We define speed-eating as a fast rate of movement or action when young people put food into the mouth, chew and swallow, in order to finish their food as fast as possible. This can be interpreted as an attempt to escape from parental and teacher control at mealtimes," I read in an article exploring "the realm of food consumption practices as a political arena."[10] My kids were demon speed-eaters, but I interpreted it more as an attempt to escape back to messaging, Facebook, and Dune.

The Experiment proved this to be a very powerful hypothesis.

With no more attractive prospect to lure them from the dinner table, the children did not exactly learn to linger over cigars and brandy. But at least they stopped inhaling their food and bolting for the nearest digital foxhole. We did slow down, all of us, and, over time, we did engage in more meaningful dinnertime dialogue. But then, given our prevailing standard—"So how was school?" "What?" or "Why aren't you eating your peas?" "What?" or "What's the deal with the Carbon Trading Scheme?" "Who?"—that wasn't hard. Overall, I would estimate that we probably increased our face time at the dinner table by 15 to 20 percent, in both quantity and quality. That was pretty good, I guess—but still less than I'd expected. I'd pictured us like something out of a Norman Rockwell painting, engaging in spirited but civilized debate, our faces aglow with family feeling and an excess of giblet gravy. The truth was, we were still more likely to bicker over who got the Hannah Montana glass.

There were unexpected gains elsewhere on the bill of fare. Deprived of his early-morning downloads, Bill started spending more time at the breakfast table. He didn't initiate a lot of conversation. But he did eat a lot more eggs, and spent an impressive amount of time reading the sports pages. I'm not sure it improved family communication, but it made me smile to see him tented importantly behind the pages of *The Australian*, like somebody's father. Sussy, too, eventually started to make unscheduled appearances at the breakfast table. "Do you want oatmeal?" I'd ask. "Eggs? Toast? Juice? A smoothie?"

"No, thanks," she'd croak gruffly, gulping her tea. Then I'd serve whatever it was I was making for Bill, anyway, and she'd eat every bite. It was sort of the opposite of demand feeding—more supply feeding, really—and I wished I'd started it fourteen years earlier.

At the most basic level, The Experiment forced us to notice food more—just as we noticed music more, and sleep, and each other. Before, eating had been a side dish. Now it was the main course, or at least one of them.

Our approach to cooking changed too, especially for the girls. They'd started out as reasonably competent cooks, but by the end of The Experiment they were capable of turning out entire meals with ease. More important, they *wanted* to. Bill, alas, responded by growing even lazier in the kitchen. This was especially true once he got a job and enough pocket money to supply his bubble-tea habit. On the other hand, The Experiment did ignite his interest in the barbecue, in the true, albeit slightly cringeworthy tradition of the Aussie male.

Our shopping habits morphed in intriguing and unanticipated ways too. Before, I'd often shopped for groceries here and there, dashing up to the supermarket or deli on a need-to-nosh basis. Now, the Saturday morning shopping trip became an essential weekend ritual. Pre-Experiment, I'd always shopped alone. Now Anni came with me, eager to help plan meals and to steer me tactfully toward more adventurous choices of yogurt and cookies. (I am the sort of person who can, and in fact has, eaten the same brand of chocolate chips for twenty-three years.)

Once the reality of the global economic downturn started to bite, we determined to become better recession shoppers, and the child who once told me she'd "never actually been hungry" even got interested in planning meals. The chore of shopping for a family became more palatable, less of a burden and more an event—even an opportunity to bond. It also made me aware of how incredibly rigid my grocery choices had become. I found myself taking daring steps. Buying dishwashing liquid with the passion fruit scent, or paper towels stamped with *different* unidentifiable pictures. And who could forget the egg-ring incident?

"Wow. Egg rings! I've always wanted egg rings!" I sighed as we passed a display in the kitchen gadgetry aisle.

"You say that like it's some impossible dream, Mum. They're two ninety-five, for crying out loud. Just buy them."

The lunacy of it all cracked us up. By the time I had the strength to reach out and take a packet, our faces were streaked with mascara and shoppers were giving us a wide berth. For the next week, we enjoyed a festival of unnaturally round fried eggs. It was as I'd always imagined. They *do* taste better.

To him whose elastic and vigorous thought keeps pace with the sun, the day is a perpetual morning.

—*Walden*, chapter 2

As our overall rate of blobbiness declined—in our eating habits, in our eggs—we also started to firm up the boundaries between night and day, sleeping and wakefulness. At the outset, giving up the 24/7 lifestyle that had featured cruising for eBay bargains at midnight, posting status updates at four a.m., and sleeping with phones under our pillows "just in case" (of what? falling finally into REM sleep?) was a rude awakening for the whole family. Yet it soon became apparent that the less we used our technology to "chill," the more rest and sleep we enjoyed. For me, that was a real wake-up call.

It's not as if I didn't know that sleep was important to the way we function. The dawn of every day brings a new study about the dangers of "sleep debt." Like most educated parents, I was well aware of the alarming [*sic*] evidence that most of us are getting far less sleep than experts tell us we need. It was the direct link we experienced between sleep (or lack thereof) and technology (or lack thereof) that started to sound in my head like a gong. In fact, plugging back into our diurnal rhythms—getting more and better sleep each night—arguably had a greater impact on the quality of our lives and relationships than any other single factor during The Experiment.

In Sussy's case, "sleep debt" was too weak a phrase. "Sleep

bankruptcy" was more like it. When she'd started fifth grade at a pricey private school—think tartan pleats and hair-extension-destroying berets—and received her own MacBook as part of the school's laptop program, I'd watched her confidence grow like corn in the night. Bit by bit, her dependence on the laptop did too. The "learning aid" that was genuinely helping her to be a more creative and more productive learner by day was opening a totally different can of worms as soon as she left the classroom. In the months leading up to The Experiment, she was spending virtually every waking after-school hour in its company: slumped on her bed, her fingers flying as she flitted between half a dozen windows of MySpace and Instant Messenger, pausing to check the done-ness of an assortment of music or video downloads.

I'd wake at two or three in the morning, channeling *Madeline*'s Miss Clavel ("Something is not right!") and stumble down the hallway to behold my baby still astare, often in full school uniform, her eyes as wide and glassy as DVDs. Most of the time she surrendered the device wordlessly, though whether from exhaustion or obedience it was impossible to tell. I had noted that her resignation as she handed over the goods seemed mingled with relief, like a five-year-old playing with matches—fascinated but at some level frightened. Half wanting to be caught.

When we first talked about The Experiment, Sussy warned me she wouldn't be able to sleep at all without her laptop. I pointed out that she wasn't able to sleep at all with it. "It's not like I'm not trying," she insisted. "I just . . . don't . . . sleep." I was reminded of the time some years earlier when she declared herself unable to smile for a family photograph. "It's not my fault," she wailed pathetically. "I've forgotten how!"

Our Wi-Fi didn't extend as far as Bill's room—thank the Lord for low signal strength!—but Anni's room, like Sussy's, had a sweet spot if you angled your laptop just right (and trust me, she did, even if it meant hanging sideways off the top bunk). Anni had gone as a scholarship student to the same laptop-centric school Sussy now attended, and she'd been equally vulnerable to the siren song of 24/7 connectivity. The difference was, then there'd been so much less to connect *to*. The four years between the girls had seen the dawning of the golden age of Web 2.0. The ensuing rise of interactivity and social-networking utilities had turned the Internet from a glorified public library to a transglobal theme park. Forget about heading to the malt shop after school. With the appearance of MySpace the world was their malt shop. It never closed, and it never stopped serving.

Bill, who was in public school, had never acquired a laptop. His nighttime ritual, pre-Experiment, was all about television. He had the monster set in his room—the only child in our family who enjoyed that dubious privilege—and formed the habit of falling asleep to its comforting flicker and drone. Even if the girls were watching the same show in the family room, Bill preferred to watch his own set—occasionally making a cameo appearance during commercials. Lord knows why he was so attached to it. As described, it was old, and the reception was laughable, despite—or possibly because of—a primitive set of rabbit ears he'd rigged up out of a wire coat hanger. Most mornings when I'd go in to wake him for school, I'd find the television still on, its hectic presence dominating the tiny bedroom like graffiti that could talk back.

During the midterm interview—in the bleak midwinter of our disconnect—Bill was as uninformative as . . . well, as a fifteen-year-old boy being interviewed by his mother. I was not overly concerned. The impact of the previous three months had been so obvious in Bill's case, I was really only going through the motions. "Do you miss TV?" I asked him almost as mechanically as he'd been answering.

"Not really," he grunted. And then, just as I was ready to move on from this stunning adventure in self-analysis, he added a postscript: "Anyway, I sleep better." When I pressed him for details, as a mother does, he went all evasive on me again, as a son does. He insisted the phenomenon couldn't be described. "How can I describe it? I don't understand the biomechanics of sleep!" he protested—a deliberate misreading of my desire to know more. "I just feel more refreshed after my sleep, okay?"

Because Bill had always been such a reliable sleeper, generally dropping off (in his own words) "like a man falling into a sewer," it had never occurred to me to worry about the quality of his sleep. Later, when I read that the American Academy of Pediatrics recommends parents remove television sets from their children's bedrooms, I realized I'd been dreaming.

As we have already noted, a bedroom television predisposes kids to eat more junk food, read less, and—most obvious of all—watch more TV: four to five hours more a week. The 2010 Kaiser Family Foundation study found "more and more media are migrating to young people's bedrooms," with more than three-quarters of eleven- to eighteen-year-olds owning a personal TV and a third enjoying bedside Internet access. Twenty-nine percent of Americans eight to eighteen years old own a laptop: the ultimate in wall-to-wall media convenience and increasingly the device du jour for watching video content anyhow. The evidence is incontrovertible: The more time kids spend "screening," the less time they spend sleeping. Less obviously, the relationship between family meals and sleep is highly correlated too: The more time children spend at family meals, the more time they spend asleep (presumably not simultaneously), according to research published in the *Journal of Family Psychology* in 2007.

When researchers Steven Eggermont and Jan Van den Bulck examined the use of media as a sleep aid among 2,500 teenagers, they found that more than a third watched TV to help them fall asleep; 60 percent reported listening to music; half read books, and more than a quarter of boys—but only half as many girls—played computer games. Across the board, teens who fell asleep to music, TV, or the computer screen "slept fewer hours and were significantly more tired" than those who read or used no sleep aid.[11] The real question—apart from "How the hell can anybody fall asleep playing a computer game?"—is *Why?*

No one knows for certain why screen-based media seem to wreak havoc with children's sleep patterns in a way that reading doesn't. But one hypothesis is that the bright light emanating from computer, TV, or even MP3 screens may interfere with the release of melatonin, a naturally occurring hormone important in the regulation of circadian rhythms.[12] The so-called hormone of darkness, melatonin is normally secreted by the pineal gland in the middle of the night, but exposure to light can significantly reduce melatonin levels, which in turn disturbs the sleep-wake cycle. There is also a link, albeit less understood, between melatonin and immune function.

A Finnish study of more than seven thousand children aged twelve to eighteen found that intensive media usage was associated with poor perceived health, especially (or in some cases only) when kids' use of technology was interfering with their sleep. Not surprisingly, there was also a correlation with increased daytime tiredness. Among older teens, researchers noted a clear gender divide, with boys most at risk from intensive computer usage, and girls from overuse of cell phones.[13]

What's so important about sleep in the first place? Researchers are only just waking up to the facts themselves, now that the entire developed world is staggering under an unprecedented burden of sleep debt. Recent surveys show that about one-fifth of adults report insufficient sleep. Among teens, the figures are even worse, with as many as a quarter clocking in six hours or less a night, compared with a recommended minimum of nine hours for their age group. Disturbed sleep is associated with a nightmarish range of psychological, social, and physical problems. Teenagers who sleep poorly report more depression, anxiety, hostility, and attention problems. They also struggle more at school, and they are at greater risk of drug and alcohol abuse. Physically, they are more fatigued, less energetic, and more prone to headaches, stomachaches, and backaches.

The effects of insomnia have been widely studied. The long-term consequences of what researchers call "short sleep"—the real epidemic among our Digital Natives—are less well understood, but are believed to have even broader negative consequences for later functioning involving somatic health, interpersonal relationships, and even general life satisfaction, according to a wide-ranging review of the current literature published in 2009 in the respected *Journal of Adolescence*.[14]

Yet if it's true, as one recent study found, that only 17.2 percent of young people are actually getting the sleep that experts insist they need, isn't some degree of sleep deprivation . . . and I hesitate to use this term to describe teenagers . . . normal? Absolutely, note researchers from the University of Texas Health Science Center. "Sleep deprivation among adolescents appears to be, in some respects, the norm rather than the exception in contemporary society."[15] For once, when our kids tell us, "But Mum, everybody else is doing it!" they are making a statistically accurate observation.

Nonetheless, when you dig a little deeper, it's clear that "normal" is not the same as "the norm"—and neither term necessarily implies "healthy" or "recommended practice." A generation ago, it was "the norm" for adults to smoke in the car with small children present. Seat belts in those same cars, let alone car seats, were *not* "the norm." Now we know better, and we wish they had. Traveling on a plane recently, I watched a *Mad Men* episode in which an attractive, affluent family eats lunch in a wooded picnic area, circa 1963. When they finished, the impeccably groomed young mother simply shook their rubbish onto the lawn, folded up the blanket, and drove away. It seemed such a heavy-handed period detail—surely no one ever *did* that? But I am old enough to know that they did. I remember when the anti-littering slogan "Keep America Beautiful" seemed downright radical. In fact, I swear I remember when the verb "litter" appeared for the first time, in 1960. (Yes, okay. I admit it. I Googled it.)

The "normal" but still mind-blowingly destructive sleep patterns we tolerate today are the result of a huge range of changes in the way we live, from the long-hours lifestyle our jobs demand (or encourage) to the trend toward smaller families and more relaxed household rules. (Behavior that would be untenable in a bigger brood—kids going to bed when they feel like it, for example—can be accommodated when there is only one child at home, or even two widely spaced ones.) Technology is therefore not to blame for our global sleep debt. But the role that it plays in extending and entrenching those dysfunctional patterns within families is significant and, especially as far as our teenagers are concerned, alarming.

I observed it firsthand in my own household, as we wound back the clock to simulate a simpler and yes, sleepier, era.

When Sussy returned home in mid-February and surrendered her laptop, she was more square-eyed and exhausted than I'd ever seen her. The six-week stay of execution at her dad's had been fun, she reported, but also a little lonely. She missed the chaos of kids coming and going, of the pets' annoying but adorable attention-seeking. And Hazel the Handheld Kitten was a powerful draw card. She may even have missed having her mother breathing down her neck, who knows? (Although she maintained that her father and I were equally strict, only about different things—probably a fair call.)

Her first reaction to the sensory deprivation tank that we now called our home was to flop heavily on her bed and lose consciousness. This didn't surprise me much. Before The Experiment, Sussy catnapped like a newborn or a narcoleptic. She napped after school. During school (or so I suspected). Even in the morning, after she'd put on her uniform, when she'd clump back into bed with her black lace-up Oxfords. Or she'd sleep through all of Saturday, waking up at four or five p.m., bright as a button and agitating for a sleepover. Come Monday morning, I'd practically need a forklift to drag her out of bed. She also had a knack for what I thought of as "defensive napping." If there was time to kill before a big event, or an obligation she preferred to avoid, she could somehow will herself into an accommodating slumber. On Christmas Eve, while the other kids lay wide awake for hours with visions of sugarplums dancing in their heads, Suss went out like a blown fuse. "How do you *do* that?" we'd all beg to know. She'd smile mysteriously. "I just tell myself." was the only explanation she'd give us. It was all the more magical given that, on a normal night, she was capable of fussing and fighting sleep for hours. "Just friggin' TELL yourself." I was often tempted to shriek.

Now, recovering from a nonelective laptop-ectomy, she channeled her extraordinary capacity for inner hypnotic suggestion once more. But this time, it went way beyond the usual defensive napping. Not only did she sleep until noon, she slept until noon having gone to bed at 7:30 p.m. And she did that not once or twice, but on and off for a month. On the weekends, she was in virtual hibernation, emerging like a bewildered bear cub to seek sustenance (usually at odd hours), or to call Maddi on the landline. Despite this, she also missed some school, pleading tiredness. I tried to discourage this, but there were days when I didn't have the heart. She seemed so listless, so wrung out.

Under other circumstances, I would have had her assessed for clinical depression. As it was, I decided to bide my time, choosing to frame it as a withdrawal, not a clinical mood disorder. It was comforting to see that when Suss was awake—admittedly a rare occurrence—she was cheerful. She also stayed in touch with friends after school and her appetite seemed fine.

Equally important, while the behavior was slightly pathological—or at least way beyond the norm—it also made sense. At one level, it was just another manifestation of the ingenious avoidance tactic she'd perfected years ago. The prospect of life without screens was simply too unpleasant, or perhaps too confusing, for her to confront head-on with full conscious awareness. But I also became convinced that her oversleeping was a way for her body to literally make up for lost time—to pay off a sleep debt that had been compounding menacingly for years.

Well, that was my theory. And when, at length, she awoke from her torpor by around week five, I'll admit I felt vindicated. It would be an exaggeration to say Sussy arose like Snow White (albeit sporting hair extensions and DIY fake tan), the spell broken forever. There were no bluebirds twittering from her shoulders on a school morning, trust me. But something did, rather suddenly, switch on for her—or off again. On school nights, she started going to bed before eleven, and within another week or two by ten. She allowed herself to be roused by seven—and I no longer had to play my part of human snooze button to make it happen. Most mornings, there was time for breakfast. Sometimes—and here's where it gets really weird—eaten at the table in the kitchen. On the weekends, instead of sleeping in until noon, one, or even two p.m., she woke up at around 9:30 a.m. "Random!" she cried the first couple of times it happened. "What is *wrong* with me?"

The Experiment confirmed my strong suspicion that media had been robbing Sussy of sleep for years. She'd been our family's most militant multitasker, and the one who'd gravitated to a digital lifestyle at the youngest age. Unplugged, the changes to her sleep patterns, energy levels, and mood were correspondingly dramatic.

The evidence strongly suggests she is no isolated case. For Generation M, the links between our diurnal habits and our digital ones are as direct as they are disturbing. Add this to the already crumbling boundaries around family diet, and the synergy is unmistakable.

A 2009 study of one hundred Philadelphia-area children aged twelve to eighteen, published in the journal of the *American Academy of Pediatrics*—the same respected body that has recommended banning TV from children's bedrooms—found that kids who spend more time online also drink more caffeinated beverages, with a resulting one-two punch to their prospects of good sleep hygiene. "Subjects who slept the least also multitasked the most," the authors concluded succinctly.[16] Among heavy multitaskers, more than a third took naps after school; 42 percent did so on the weekend; and a third reported nodding off at least twice a day. One child in the study, who slept for an average of five hours a night, reported falling asleep eight times during a typical school day. In the opinion of the researchers, the upshot was a recipe for "changes in school performance, difficulties with executive function, and degradation of neurobehavioral function."[17]

Among American teenagers, sleep duration has decreased by one to two hours over the past forty years. The proportion of kids who sleep for fewer than seven hours a night has doubled in that time.[18] To address the problem, many educators and other experts have recently started calling for a change in school hours, arguing that they are a poor fit for teenagers' "natural biological rhythms." That may well be. But no one seemed to worry about that forty years ago, and school hours were exactly the same as they are now. If our Digital Natives are in danger of becoming a generation of sleepwalkers—and the evidence suggests that is exactly the direction they are tending—maybe we could *all* use a wake-up call.

May 21, 2009

Christian, German exchange student friend of B.'s, here for weekend. Apologized about lack of technological hospitality but he seems more intrigued than repelled. ("My mother alzo vants to do zis!") Boys amused selves taking Rupert to the beach, playing music, hanging out in front of fire, taking 2.5-hour nap (!!). Headed out to a party by bus after dinner.

A. & S. prognosticating at dinner on the subject of their future careers—a possible joint business venture of an indeterminate nature. They imagine "hell beautiful suits," excellent hair ("I see a bun . . ."), and very large, very light-filled, very white-on-white offices.

"What will you actually be doing, though?" I ask. They look at each other and shrug. "No idea!"

May 22

B. delivered drunk at midnight, vomiting into the bougainvillea, croaking apologies as I stalk off to bed. Disgusted by the spectacle. Pleased, perversely, by the awfulness . . . assuming in my ever-hopeful way that there is such a thing as a lesson you don't forget.

Discovered on couch at seven next morning, gravely reading Murakami (still reeking even after shower . . .). It was dark rum, he says, swigged straight from the bottle. Barf me out.

May 23

S. & A. fighting like catfish (do catfish fight?) over kitchen mess. If more cooking equals more mud-wrestling, takeout is looking good.

Mollified by news of S.'s top mark in math test. "This studying thing . . . I'm thinking I might try it again sometime," she muses. We celebrate by eating cupcakes and inspecting Hazel's teeth, exclaiming over their tiny perfection. Handheld kitten that she is, she endures patiently.

Okay, it's not *America's Got Talent*, but it works for us.

May 27

S. home sick with stuffy nose, fever. ("P.S. I feel like I'm going to throw up.") She and Maddi conferring already on festivities for Independence Day, aka Return to Screens Day, on July 4. Should we make a video? What about B.? He'll be in Germany on his water-polo tour then. Can we Skype him?

B. rehearsing for gig at local pub. Learning Miles Davis's "Blue in Green." Soaring, soulful, effortless, lyrical.

Thrilling-slash-annoying how much time now spent playing Satie (and/or humming ostentatiously to Brandenburg Concerto score—they are learning it in Music) while girls and self play Boggle. Hazel's dentition continues to enthrall.

May 29

B.'s first gig! More a giant public jam session—twelve onstage (including B.'s teacher, who helped organize), eight in audience—but music good, and B. marvelous in school shorts and his orange Nikes.

May 30

To movies to see *My Year Without Sex*—charming/biting in equal measure. Favorite scene: kids at dinner party grumbling when all forced to watch the same DVD (despite being plugged into iPods, phones, and games anyhow). Says one prepubescent, rolling her eyes, "I can't believe you only have ONE TV!"

S. summoned home from sleepover to clean room. Arrived scowling with friend Sean in tow, who lolled like a sultan on couch in

TV-less TV room, glaring into the corner where set once stood. Willing it to reappear, maybe.

Experimental conditions such a litmus test for friends . . . Really separates digital sheep from digital goats. Sean clearly among former. Ten minutes screen-free and he's twitching like he's spent a month in Gitmo.

May 31

Arrived home at nine p.m. to discover B. and four friends cooking eighteen fish on the grill. An impromptu birthday thing for Vinny, they explain. And a herring run at South Beach. I look around. The grill is still hot. Garlic and lemons litter the table. My jaw drops like a cod. "You . . . can . . . barbecue?" Shrugs.

Normally, I tell boys, he struggles to add flavor packet to instant noodles. They laugh conspiratorially.

June 1

Generally teary and emotionally labile all day. Injured no one appreciated my excellent chicken-mushroom stir-fry. Fought with B. on way to water polo. Laughed till I cried while writing column—then sobbed hysterically over awful news headlines. S. played me some hokey Taylor Swift song and I cried some more. After dinner, settled down to read *Five People You Meet in Heaven*—meant to be deeply inspirational and uplifting—and howled all the way through that too.

And so, soddenly, to bed.

June 2

Forced by B. to do the math, his friend Patrick calculates has spent six hundred hours playing Task Force 2 in past twelve months (equivalent to four months' full-time work). "And he says he has no time to practice drums!" B. snorts.

"Hang on!" he cries in genuine horror. "That sounds like something *you* would say!"

Mary and Grant and family over for dinner prepared entirely by A.—assorted appetizers, amazing lasagna, salad, chocolate slice. Kids ate separately at cleared, candlelit craft table, followed by raucous game of Articulate and three hours of good, old-fashioned face time.

June 3

Was convinced my bank account had been skimmed, then that my briefcase with Filofax had been stolen from car. Mistaken on both (and innumerable other) counts.

Menopausal much?

B. reading six hundred–plus page Murakami novel on couch after dinner, wrapped in blanket like large purple burrito, Cannonball Adderley playing "Somethin' Else" (recorded the year I was born). Fell asleep by nine with the blanket pulled over his face, and Hazel's tiny head resting on his open palm.

June 4

One month and counting: anticipation, apprehension. Ambivalence!! Signed us up today for new Internet/phone bundle to connect in exactly thirty days' time. Also made top-secret Foxtel inquiries. Yes, subscription TV, cable, a hundred channels, and nothing to watch. It seems something deep within me is longing for *television*! (Could this be my version of hot flashes?)

Girls reenacting chunks of *Stepbrothers* for me—"We are censoring it, Mum, don't worry!"—then making hot cocoa and massaging each other's feet.

A. invites us to smell her new fake tan. We accept. I serve stewed apples with whipped cream, which S. describes as "the best thing I've ever eaten in my whole life." We sing along to Miley Cyrus/Taylor Swift mix tape—shame!—and dance with pug.

June 5

Long conversation with A. after dinner about the *nature of evil*. (Was it something about the meatloaf, I wonder?) Six months ago, was lucky to delve into the nature of bad hair days with her. Is reading *Too Soon Old, Too Late Smart* (at my suggestion) and swallowing it whole. Copying quotes into her journal.

Random! We are now a three-journal family.

June 6

After dinner, S. to Andreas's house around the corner—a boy she hasn't spoken to since Year 4—to, quote, "use his Internet!" No punches pulled there. Back at 9:30.

Visited Bill at new job—waiter and kitchen hand at Tasty Express—for muffin and coffee he barista-ed himself. Revelatory to see him as a public figure, dressed in a uniform and serving.

A. drove expertly to B.'s gig tonight—parking too. (Said proudly with straight face: "My instructor says I reverse even better than I drive!")

June 8

Found myself leafing almost tearfully through a Radio Shack circular this morning.

Apple strudel yesterday, butter cake with orange glaze today. Let us eat cakes!

June 10

Everybody over it now. All a bit sick, a bit cold, a bit bored, a bit ready to curl up for the winter with a screen—*any* screen.

Girls propose getting a new black pug to celebrate Back from Black Day. I'm thinking black sofas. And a new coffee table. And one of those entertainment units that look like a tabernacle to television.

Amused ourselves after dinner going through a tin full of kids' old cards and "vouchers" (like Bill's to me, aged nine, promising to "be sensible for half an hour," and Sussy's Mother's Day card announcing her love was giving her a "hard-a-tack").

Then fudge, then tea, then Boggle, then bed.

June 20

IKEA today. Allen key repetitive strain injury tomorrow.

Pleased nonetheless with sofas, tables, and a zillion new lights (as in "let there be . . .") for TV-room rebirth.

Confiscated S.'s phone just now as has been soldered to it all weekend, along with new cordless phones—yeah, I caved—which she is deconstructing feature by feature.

Reclaimed my study—aka the family phone booth—with a huge clean/wash/vacuum. Now that we are cordless again, have declared it a teenager-free zone once more.

June 21

Transformation! As if the family room has finally burst from its chrysalis: full of warmth and comfort and light after its Lenten gloom.

Kids arrive home as I screw in final frigging bolt. So like Christmas morning (to mix a holiday metaphor), everyone oohing and aahing and throwing themselves onto squishy new sofas face-first.

Euphoria snuffed out when B. asked about The Beast, and I told him I planned to relocate it to the passageway outside my study, out of sight lines of the television. Ugly exchange ensued but held my ground.

Ten minutes later we were debating with equal intensity—wait for it—the existence of God. It started with S. asking if I'd ever heard of Pascal's Wager and B. entering the fray with full-throttle Dawkinsian ferocity and A. shouting out her two cents from the bathroom.

Note to self: Atheism a much better outlet for sixteen-year-old rebellious energy than furniture arrangement.

Before bedtime S. told me she's decided to take down her MySpace page. Explained, "I guess I'm just over stalking people."

"Hey, there's something you can put in the book," she added proudly.

The Return of the Digital Native

He said, about leaving Walden . . . "Perhaps it seemed to me that I had several more lives to live, and I could not spare any more time for that one."

—AIDEN402, *Best Answer to Yahoo! Answers question "Why Did Thoreau Leave Walden Pond?" http://answers.yahoo.com/question/index?qid=20080208052927 AApsk2y*

I learned this, at least, by my experiment: that if one advances confidently in the direction of his dreams, and endeavors to live the life which he has imagined, he will meet with a success unexpected in common hours.

—WALDEN, *chapter 18*

It was a shocking story no matter which way you looked at it, so maybe I shouldn't have been surprised by Sussy's distress when I recounted the details. How a stressed single father is in the kitchen preparing dinner for his three boys. How he calls the boys to the table but, as with the proverbial tree falling in the forest, no one hears him. How he calls again. And then a third time.

"Finally," I say, "the father charges into the family room—and there the boys are, oblivious in front of their huge plasma TV. They are multitasking, of course, and have earphones in, so when he shouts they *still* don't hear him. That's when he loses it."

"He makes a lunge for the television and hurls it right through the sliding glass door, where it shatters into a million pieces on the deck."

Sussy swallows hard. "And then what?" she asks in a little voice.

"Well, the neighbors hear all the noise, of course, and call the police, and they file a report. And in the end the Family Court gets hold of it, and, well, the man *loses* his children."

It's a pretty full-on story, and now I'm regretting I've told it. I can see Sussy is on the verge of tears. "You mean . . . you mean . . ." she quavers, "they never get the plasma back?"

Let's face it. From a Digital Native's perspective, pulling the plug on a person's screens is pretty much pulling the plug on life itself. When I think about that, and think about the last six months, I am full of pride for what my children have accomplished . . . or endured . . . or whatever you want to call it. I'm pretty sure they feel the same way.

In the United States, high school students reading Neil Postman's *Amusing Ourselves to Death*—the classic work on television culture— are advised to undergo a twenty-four-hour media fast to road test the ideas for themselves. Most are gutted by the effort. When I tell the children this at dinner one night, their collective scorn is wonderful to behold. "Pussies!" Bill mutters.

On the eve of Independence Day 2009, it occurs to me that I should sit everybody down for one final interview. Or maybe I should assign an essay topic ("What I Did on My Holiday from Technology"). But the truth is, I already know the results of The Experiment. We all do. Like Barry Marshall swallowing that vial of ulcer-causing bacteria, we have been our own observers, and our own observatory, the whole way through: researchers, subjects, and peer-review panel all rolled into one. There are no surprises at the end. And there is nothing particularly subtle about our findings either. The hypothesis (or did it start out only as a hope?) that six months without screens would cause us to reconnect with "life itself"—binding us together as a family, propelling us outward and upward as individuals—has been confirmed so often and in so many ways.

Collectively, we have stared down the Gorgon of boredom and learned to find diversion in the unlikeliest places: in the morning shadows on a bedroom wall, inside a kitten's delicate maw, in the suddenly revelatory details of a suburban streetscape. Unstoppered from our digital dummies, we looked around for some substitute means of soothing our spirits and found—among other things—each other. We hung out on each other's beds, and on the couch in front of the fire. We lingered for no good reason over dinner. We invaded each other's space. Whereas before we'd scurry to our separate corners, we now found excuses to band together and stay there. As a family, our talk became more interesting, and our conversations more challenging, for one simple reason: because they had to.

Sure, there were expectations that remained unfulfilled, for both good and ill. I'd anticipated that life without the soundtrack of my iPhone would be a cross between a Celtic dirge and Patsy Cline singing "Crazy." Instead, I reached the acceptance stage of grief practically unscathed. Who knew? I dreamed, too, of instituting a whole raft of sixties-inspired family rules. No eating between meals. Lights out at 10:00 p.m. None of your back talk, young man. I'd also hoped I'd lose twenty pounds, take up triathlon, develop more prominent cheekbones, and save money. None of these things, in the end, materialized either.

Then there were the outcomes that hadn't been reckoned into the game plan at all. And these, possibly *because* they were so unexpected, were the most memorable of all. Bill's metamorphosis as a musician, to take the most obvious case in point. At the time, it seemed such a bolt from the blue. But in retrospect I realize it was an evolution that unfolded logically as well as lyrically. Like an improvised solo that blows you away with its effortless, in-the-moment perfection, artfully concealing the hours of work it took to get there.

Bill remarked at the time that The Experiment was only a trigger for his musical awakening. I see the process in more *Walden*esque terms, as a pebble dropped in a pond. Time freed up during The Experiment forced Bill to put down his gun, virtually speaking, and that created the first ripple. Giving up gaming didn't simply make a space for new growth. It created a massive crater, a vacuum all but shrieking to be filled. The second ripple was all about reconnecting with a different group of friends. Tom, Matt, and Will were already marching to the beat of a different drum kit, to paraphrase Thoreau's famed metaphor. Through them, and his new teacher, Bill was introduced to a massive, multiplayer game with an infinite number of levels. Jazz.

Sussy's six months of detox worked wonders on her dysfunctional media habits, just as I'd hoped. What I hadn't anticipated was that the landline would become her methadone. Okay, so win a few, lose a few. The impact on her sleep hygiene had consequences both expected and surprising. In the months leading up to The Experiment, the baby had definitely hit rock bottom as far as quality and quantity of sleep were concerned. A spell in digital rehab was exactly what the doctor would have ordered (though he might have had to friend her first on Facebook).

What I hadn't anticipated was the ripple thing. I didn't imagine that getting more sleep would create a kick-on effect in virtually every aspect of Sussy's life and would transform her relationship with me and her siblings. Her moodiness and all-round bolshiness, which we'd put down to being a teenage girl (increasingly spoken of as if it were a preexisting condition) turned out to be largely a function of being *tired*. The Experiment had given us all a gift of time. But to Sussy it was literally a gift that kept on giving. With her sleep debt finally fully paid up, with interest, she was up for so many new things—cooking, reading, beating the bejesus out of me in Boggle—because she now had the *energy* for new things.

Anni, as the eldest and most independent, suffered the least inconvenience and probably underwent the fewest changes; paradoxically, the process of unplugging seemed to give her the most pleasure. After her initial fit of pique, and pacified by her "incentive" (as I preferred to think of the cash bribe I'd offered), Anni more than any other child consistently supported the enterprise. In fact, she'd prefigured it, having spoken in November 2008 of a plan to undergo a self-imposed Facebook fast for the holidays. Over the course of the six months, she achieved a number of important milestones, including completing two demanding professional journalism internships. (Her unflinching reporting for a magazine feature on Perth's private girls' schools, "Lock Up Your Daughters!," sounded alarm bells throughout our community.) She also passed her driving test, logging in her required supervised hours with her mother clutching the parking brake and wearing a bald patch into the passenger-side floor mat, to finally obtain her license.

Would Anni have cleared these hurdles if The Experiment had never happened? In all likelihood, yes. Would the wheels of change have spun slower, wobbled more widely, and required more frequent roadside assistance? You betcha. The sheer expanse of liberated time—in Anni's case, upward of thirty-five hours a week—made a mockery of her usual procrastination strategies. More important still, the digital drought at home propelled her outward in search of deeper and more biologically diverse waters. Now that her social survival depended on it, connecting more directly with the challenges and gratifications of "life itself" was a no-brainer.

Our technology had ostensibly put the world and all its wonders at our fingertips. But in truth we were more like the frog in the Chinese fable—the one who lived in a shallow well but, because he had never seen the ocean, imagined himself master of the universe. The Experiment dragged us up to the sunlight and out to the shore. Without a webcam, it looked so different out there.

I had a lot of time to contemplate such matters, especially once the winter weather set in and we started gathering in front of the fire after dinner, like something out of *Little* freaking *House on the Prairie*. In *Walden*, Thoreau observes, "At length the winter set in good earnest . . . and the wind began to howl around the house as if it had not had permission to do so till then . . . I withdrew yet farther into my shell, and endeavored to keep a bright fire both within my house and within my breast." Yeah, same!

We'd begun the Winter of Our Disconnect in glorious summer. By June the winter storms began to rip up the southern coast, raining blows on our west-facing front door and causing the curtains to billow even after the windows were shut tight. That last month felt longer than the previous five put together. For me, being housebound made our screenlessness feel like a hardship for the first time. I longed to curl up with a movie, to watch *60 Minutes* or even *Australian Idol* (that's how far gone I was) on a Sunday night, to escape the confines of our insufficiently insulated four walls with my beloved NPR news ticker, or *The New York Times* online, or maybe a spot of harmless WILFing.

Instead, I brooded and Boggled and fetched more wood.

"Every man looks at his woodpile with a kind of affection," Thoreau observed. "I love to have mine before my window, and the more chips the better, to remind me of my pleasing work."

My woodpile was under the carport, but I started to feel pretty much the same way about it: which is to say, overly fond. "Dudes, check out my kindling!" I'd urge the kids whenever we passed by. The indulgent, slightly sad way they smiled was unnerving. But the truth was, watching the fire was the closest thing to live home entertainment we had left. And "watching," passively watching, I realized, was important.

I recalled that in the primordial mists of the Screen Age, critics frequently referred to television as the "electronic hearth." I got that now, because for me the hearth had become a kind of combustible television. Seeing us all gathered around its glow—our eyes vacant, our jaws slack, our critical faculties set to sleep mode—you would have thought we really *were* watching *Australian Idol*.

What was going on in the children's heads, I am rather relieved to say, I have no idea. For myself, though, these long winter evenings were all about pondering mistakes of the past (a topic no longer deflectable with massive doses of new data) and plotting new and improved mistakes for the future. (LOL.)

Weirdly, The Experiment, for all the powerlessness it had conferred, had given each of us a renewed sense of agency. Maybe it was something about being forced to "make our own fun" that prompted it. The same Thoreauvian spirit of self-reliance that inspired us to create our own entertainment—whether it was making cupcakes or conversation, composing music or a new family ritual—spilled over into the shadows too. Staring night after night into the fire, with no plotlines except my own to distract me, I started to reflect deeply on other aspects of my life in which I'd allowed myself to be a passive receiver. Where I'd failed to "see choice"—or, to put the same thing a different way, where I'd *chosen* not to see choice.

Living in Western Australia, to take the most obvious example. There was a time in my life when it was more or less true that I'd had "no choice" about this decision. Back when my marriage ended, when my children were young and vulnerable—not to mention my career prospects—my only real choice was *how* we'd live here, not whether we would. It now became clearer to me that I'd chosen to pour energy into making it work: caring for my amazing children, making wonderful friends, becoming part of a community I cared about, creating a home, extending myself professionally, seeking to "blend" our lives into a new and stable family unit. I'd tried with my whole heart to put down roots in this, to me, alien soil, and in lots of ways I'd succeeded. But the truth was, those roots never entirely took. They went down only so far and no farther. In recent years, instead of feeling more and more grounded, I'd started to feel increasingly vertiginous. My bio on *The Australian*'s website joked, "Maushart has lived in Western Australia since 1986, but insists she is only passing through." The last time I'd seen it, it struck me as less amusing than sad.

Adding further fuel to the fire that winter was a book called *Who's Your City?* by urban theorist Richard Florida. Its thesis—that where you live has a huge (and hugely underrated) impact on life satisfaction—was exactly the message I needed to hear. Or maybe wanted to hear, who knows? Among other things, Florida argues that cities have personalities, just as people do, and that finding the right place to live is akin to finding the right partner to live with. There are good matches, and there are mismatches—and in a world in which affluence and mobility converge to make choice possible, it is an individual's privilege—nay, responsibility—to get it right.

I started by searching the index for references to Perth. (There were several, including one study that had found it to be among the world's most neurotic cities.) Then I devoured the book whole, like one of Sussy's chocolate cupcakes. I took notes. I highlighted and underlined and wrote in the margins. In the end, I was so worked up I decided I had to take things a step further. I'd have to speak to Richard Florida directly. Working at ABC enabled me to frame my request as an "interview."

I did record it in ABC's Perth studios, from Florida's office at the University of Toronto. I even used it as part of a later program segment. But really, my conversation with Florida was a private consultation . . . and I blurted out as much midway through. I didn't need a therapist, I assured him. But an oracle would be nice. What did he think? Should I move back to New York, or stay in Perth? Miraculously, Florida was unfazed by the abrupt change of direction. Maybe he gets this a lot. Or maybe as a New Jerseyite who now lives in Canada, he could relate. The fact that he'd actually spent a couple of weeks in Perth recently was a bonus.

"Listen to yourself!" he admonished me. "You're a New Yorker! Twenty-three years and you haven't changed your vowel sounds one iota."

It was true. My accent was as stubborn as a three-year-old.

"Face it," he continued. "Perth is a beautiful city. But it's not really your city—is it?"

"Well, no. I mean, yes. I mean, I know that," I found myself confessing incoherently to this total stranger. "But I'm so afraid if I move, I'll miss it."

"Don't be afraid of that," came the answer. (Oh, this is what I'd been waiting to hear!)

"Of *course* you'll miss it."

Why do we need help in seeing what is glaringly obvious? I reflected later that night as I sat before the fire, the inert bodies of assorted drowsing animals, human and otherwise, scattered around the room like bolsters. "We don't know who discovered water, but it wasn't a fish," right?

Looking back, I realize my conversation with Florida was a turning point. Yet if anybody had told me then that within four months I'd have sold our family home, bought a turn-of-the-century farmhouse on the eastern tip of Long Island, and begun planning our transhemispheric repatriation, I'd have suspected smoke inhalation.

The official Back from Black celebration began at the stroke of midnight, Microwave Central Time, on July 4, 2009. Bill was somewhere
between Singapore and Germany that night, on tour with his water-
polo club. (Before leaving, he'd asked permission to load up his iPod
with the two trillion Miles Davis tracks he and Matt had been collecting. I'd given my blessing. I'd also unearthed the DS from the highest
drawer in the craft cupboard where I'd secreted it under a stack of
antique cassette tapes six months earlier. "Use this for good, not for
evil, son," I reminded him solemnly at the handover.) We missed him
sorely, but we knew that wherever he was, he had his earbuds in—and
was happy.

Sussy and I headed to the ballet to beguile the hours that yawned
between dinner and midnight. It was the opening night of *Romeo and
Juliet* and if we skipped the after-party and grabbed a couple of burgers on the way home, I figured we'd arrive with exactly ten minutes
to spare to put on our pajamas and jockey for couch space for the
big event. Anni was heating up our creepily beloved popcorn maker,
Poppasaurus (an orange dinosaur who obediently vomits the finished
product into a waiting bowl). Along with my stepdaughter Naomi—
the one who sold virtual real estate on Second Life and who, despite
(or possibly because of) this, had been an enthusiastic supporter of
The Experiment from Day One—she had already picked up the
DVDs whose selection we'd all been haggling over for weeks.

I'd begun the gala reconnection festivities a full month previously,
scouting deals for a new Internet service provider, browsing subscription TV packages, hiring an electrician to repair an accumulation of
blown light fittings. I'd also decided to spring for a roomful of new
furniture: two oversized sofas with removable cotton covers, a sensible and sturdy coffee table, and an "entertainment unit" ludicrously

out-of-scale to our puny television—which I dragged out of the shed exactly five months and twenty-nine days after I'd dragged it in, this time with the help of the Foxtel man. Yes, gentle reader, the Foxtel man.

Our new broadband provider offered a deal that included a VOIP phone (Voice Over Internet Protocol) that would allow us to make overseas calls for pennies. I wasn't sure how reliable it would be. But after that last phone bill, I was ready to try anything. As for The Beast, the day Bill left for Europe, it returned at last from its wanderings— dumped unceremoniously under the carport by somebody's dad, its tattered little shopping bag of travel-weary peripherals by its side. It's funny, but it looked shrunken. It was like seeing a house that loomed so large when you were a child and realizing, twenty years later, how cramped and ordinary it really was.

July 4, 2009

At 11:50 p.m. precisely, Sussy and self scream through front door and run headlong through the house locating devices, chargers, and remotes. Anni and Nome count us down as we stampede to the TV room in a spray of popcorn, Coke Zero, and pure adrenalin.

OMG. We made it!!!! And lived to tell/text/tweet the tale!!!!!!!!!!!! !!!!!!!!!!!!!!!!!!

July 5

My entry for this day reads simply "Media hangover." Well, what else did we expect? Dueling laptops, binge texting, an overdose of reality TV (*Wife Swap, Life Swap, Twenty Years Younger*) plus a DVD double feature (*The House Bunny, Superbad*) do not a restful night make. But it was the celebration we had to have, we wanted to have. Damn it, that we *deserved* to have.

Not that it was a case of unalloyed hyperconnected heaven. At about five minutes past midnight, the first technical glitch—a badly misbehaving *Stepbrothers* DVD—reminded us of something else we'd been missing over the past half-year. Frustration. (There's another good thing about books and newspapers, people. You never, *ever* have to troubleshoot them.) It brought back the hundreds of hours I'd spent waiting on a seemingly terminal help-desk queue when viruses struck, or worms bored, or signal strength faltered. I remembered the screaming matches I'd had with Bill when our broadband account was "shaped" for exceeding our monthly download allowance. "'Drugged' is more like it!" I thundered, as I waited for up to a minute—a minute! for a Web page to load. With shame, I recalled yelling words to the effect that there was no way I was going to get through the weekend at that download speed. Thinking back, it was hard to know whether to laugh or cry or just rip everything out of the wall again before it was too late.

The next day passed in a static-y blur. Blobbiness reigned so supreme it was difficult to tell which end of the day was up. "Like jet lag," as Sussy observed, with in-flight entertainment on steroids. We were sleep deprived, of course. It was 3:45 a.m. when we'd finally broken suction on our collective screens—all ten of them, counting laptops, iPods, iPhone, cell phones, camera, and TV. By ten a.m. the girls were back at their stations, watching the entire season of *Australia's Next Top Model* (which they'd downloaded overnight), and commencing to mine the educational mother lode of MTV ("Bully Beatdown," "Get Ready for Bromance," "If You Ain't Down with That"). The familiar sound of the instant message alert was heard again in the land.

Late in the afternoon, I took a long, long walk with iNez. It was such luxury to nestle my earbuds back into their rightful place, to flick to my playlists and podcasts, to encounter once more the exquisite agony of choice. Hitting "Shuffle On," there was a pause before the first song burst into the space between my ears. It was "Good Lovin'" by the Rascals, from the *Big Chill* soundtrack. Hardly a profound musical statement, but joyous—and in a funny kind of way, apt: "I got the fever, Baby, but you've got the cure . . . And I said, 'Yeah, yeah, yeah, yeah, yeah,' (Yeah, yeah, yeah, yeah, yeah) . . . Yes, indeed, all I really need . . ."

I'd be lying if I said I skipped all the way to South Beach. In fact, it was more like a tap dance.

"'Screen-time' can stunt language development and shorten kids' attention span," reported *The Australian* under the slightly hysterical headline "Ban Television for Toddlers" in October 2009.[1] In fact, the latest "Get Up and Grow" recommendations by Melbourne's Royal Children's Hospital recommend a ban on *all* screens for kids under two, including DVDs, handheld games, and computers. I'm down with that . . . in spirit, at least. But as somebody who probably has more firsthand knowledge of media bans than most, I have some serious reservations too.

There are genuine reasons to worry about the media habits of our youngest global villagers. According to figures published in *Pediatrics* in 2007—and which should therefore be regarded as conservative—the average American preschooler watches TV for an hour and twenty minutes a day. A quarter of five- to six-year-olds use a computer for another hour a day, while a fifth of under-threes, and a third of three- to six-year-olds had a TV in their bedroom. Other research has shown that the average four-month-old spends forty-four minutes a day watching TV. (Then again, the average four-month-old spends forty-four minutes a day watching her fist, so maybe that's less alarming than it sounds.) By the time that diminutive Digital Native reaches her third birthday, she will be screen-bound a minimum of three hours a day, assuming her family has a pay-TV subscription. Forget about iBrain. Nickelodeon Brain maybe more like it.

The American Academy of Pediatrics has advocated a TV ban for under-twos since 2001. Recent figures suggest that 70 percent of two-year-olds are in violation of that ban. One recent study found that children aged three to five chewed an average 3.3 hours of visual cud a day, and kids under the age of three 2.2 hours. When children were followed up at ages six and seven, and tested for cognitive development, researchers found each hour of average daily viewing before age three was associated with declines in reading, comprehension, and memory scores. And yet. And yet.

The research also shows that kids who watch more TV between ages three and five have *higher* reading scores. How does *that* work?

Australia's Get Up and Grow initiative is part of a broader anti-obesity drive that aims to encourage "activity" among the nation's children. It's not just screen-based content that's in the firing line. The guidelines define drawing, reading, and solving puzzles as forms of inactivity too. How does *that* work?

Clearly, the swings and roundabouts of kids' cognitive response to screens have yet to be fully charted. The ways in which the screen hygiene in our homes may be shaping our children's social and emotional default settings—and those of the larger family systems to which they belong—have barely begun to be reckoned.

Computer games that help kids learn to read, or add and subtract, or problem-solve are no more the equivalent of "Video Hits" than *Hamlet* is the back of the Lucky Charms box. Reading books to our toddlers—or watching television as a family—is no substitute for taking them to the park or the pool. But to imply that media are so detrimental to our children's well-being that they need to be *banned* makes SpongeBob SquarePants look like Buckminster Fuller. And if that seems like an aggravated case of the pot calling the kettle black [and white], so be it.

Media of all types—screen-based or otherwise—are as much a part of modern life as cars and planes or dishwashers and vacuum cleaners. Or, for that matter, junk food, alcohol, and tobacco. We may sometimes yearn for a simpler time or place, a safe haven completely out of range of whatever siren song we fear will lure us to our doom. Heaven knows, I understand that longing. Ultimately, however, and no matter how strongly held our convictions, we must live in *this* world—the one to which, for better or worse, we find ourselves tuned. "When one arrives at the point of reflecting about the preferability of the past to the present," observes Patricia Meyer Spacks, "it's time to change direction."[2]

As a strategy for managing the long-term media ecology in our homes, bans and blackouts are probably as effective as the Three-Day Lemon Detox Diet is for lifelong weight control. As a consciousness-raising exercise, on the other hand, extreme measures can be illuminating indeed. No amount of talk (let alone yelling) could ever have persuaded Anni, Bill, and Sussy of the extent of their media dependencies as eloquently as even a week of information abstinence. But by six months, the time had definitely come to return to what our culture (rightly or wrongly) has decided is "normal." Even Thoreau left the woods eventually.

A week or two after the Richard Florida conversation, I booked a flight home. "You dudes will have your cash bonuses to celebrate the end of The Experiment," I reminded the kids. "This trip will be my reward." What I didn't say was that it would be half holiday and half reconnaissance mission. I wasn't ready to do anything drastic. But

I felt I needed to do more than imaginatively project myself back to New York. I needed to be on the ground again.*

I left four weeks after our own Independence Day celebrations. I was nervous about leaving them, but the kids were openly ecstatic at the prospect of managing the house on their own—and with it, of course, their media—for the first time ever. "Aw, don't cry, Mum!" they called as I climbed into the cab for the airport, amid a flurry of hugs and sodden tissues. "We'll Skype you every day," Anni reminded me.

"And we'll upload photos to our Facebook pages every day too," added Sussy (obviously forgetting she'd refused to "friend" me for fear I'd become a stalker).

"Please, please don't forget the Internet phone," I begged. "It's there for a reason, guys. Use it!"

I flew directly to North Carolina to visit my parents at what I still thought of as their new house. (In fact, they'd retired to Pinehurst seventeen years earlier.) "Get it over with early," was my thinking. My expectations couldn't possibly have been lower, yet I think we were all pretty shocked at how it went. Which is to say, wonderfully and without a hitch. And with an enormous sense of . . . dare I say it? Connection. But then, as my father observed, it had been a quarter-century since we'd last spent any time together, *really* together. Saying good-bye to them felt awful and suddenly . . . I don't know, wrong. Like having a limb amputated while it was still perfectly sound.

I came up to New York on a train—a deliberate choice again. Air travel was much too abstract for my purposes. I wanted to *feel* the country, to see it trundle past at a pace I could assimilate and ponder. I wanted to sit in the dining car and drink coffee. To admire the

---

* I also intended to visit Walden Pond to pay my respects, but, alas, I never got there. "Time is but the stream I go a-fishing in," Thoreau had written. In the end, I just ran out of stream.

conductors in their handsome uniforms, making jokes as they sauntered up and down the aisles punching tickets. To watch families unpack cold chicken suppers from a basket, and hunker down for the night in their seats, with pillows brought from home.

iNez was with me the whole time, of course, as was Della, my laptop. But I never even considered plugging in or logging on. In fact, I barely read my book. Instead, I listened to the train whistle fading into the thick Southern night air, and took in the names of cities I recognized like school friends.

It was exactly as Thoreau had exhorted. Somewhere along the line, I'd become my own telegraph.

Making the move back home to New York after twenty-four years in Perth probably sounds like an abrupt change of channel. In fact, it was more the next logical step in a journey I'd been taking for years. The next ripple in the pond. Like Thoreau, we'd headed into an experiment in living that had forced us to experience life itself in a way I'd forgotten was even possible. In the end, paradoxically, it taught me that the only way forward was to go back again.

I felt I understood *Walden* as perfectly as I ever would, sitting in Starbucks on Twenty-third Street, Skyping the kids with the news of our newest experiment. "Perhaps it seemed to me that I had several more lives to live," Thoreau had reflected in the final chapter of *Walden*, "and I could not spare any more time for that one."

Whenever I tell people about our Experiment, the first thing they want to know—after "How much did you pay them?"—is "What did you learn?" If you've read this book through to the end, you already know the long answer—but just to be safe, here it is in tablet form.

The Ten (okay, Eleven) Commandments of Screen Hygiene

Thou shalt not fear boredom.
Thou shalt not "multitask" (not until thy kingdom come, thy homework be done).
Thou shalt not WILF.
Thou shalt not text and drive (or talk, or sleep).
Thou shalt keep the Sabbath a screen-free day.
Thou shalt keep thy bedroom a media-free zone.
Thou shalt not covet thy neighbor's upgrade.
Thou shalt set thy accounts to "Private."
Thou shalt bring no media to thy dinner.
Thou shalt bring no dinner to thy media.
Thou shalt love RL,* with all thy heart and all thy soul.

---

*Real Life.

INTRODUCTION

1. Donald F. Roberts, Ulla G. Foehr, and Victoria Rideout, "Generation M: Media in the Lives of 8- to 18-Year-olds," The Henry J. Kaiser Family Foundation, March 2005.

2. Victoria J. Rideout, Ulla G. Foehr, and Donald F. Roberts, "Generation M²: Media in the Lives of 8- to 18-Year-Olds," The Henry J. Kaiser Family Foundation, January 2010.

3. Sydney Jones and Susannah Fox, "Generations Online in 2009," Pew Internet & American Life Project, January 28, 2009.

4. Victoria J. Rideout, "Parents, Children, and Media," The Henry J. Kaiser Family Foundation, June 2007.

5. Alexandra Rankin Macgill, "Parents, Teens and Technology," Pew Internet & American Life Project, October 24, 2008.

6. Michael Bittman, James Mahmud Rice, and Judy Wajcman, "Appliances and Their Impact: The Ownership of Domestic Technology and Time Spent on Household Work," *British Journal of Sociology*, 55, no. 3 (2004), 401–423.

CHAPTER ONE

Who We Are, and Why We Pressed "Pause"

1. Tamar Lewin, "Parents' Role Is Narrowing Generation Gap on Campus," *The New York Times*, January 6, 2003.

2. Australian Bureau of Statistics, ABS 8153.0, Internet Activity Australia, 2008.

3. gigaom.com/2010/05/12/us-broadband-demand-bounces-back/.

4. Nielsen Wire, January 22, 2010.

CHAPTER TWO

Power Trip: The Darkness Descends

1. Tracey L. M. Kennedy et al., "Networked Families," Pew Internet & American Life Project, October 19, 2008.

2. John Palfrey and Urs Gasser, *Born Digital: Understanding the First Generation of Digital Natives* (New York: Basic Books, 2008).

CHAPTER THREE

Boredom for Beginners

1. Patricia Meyer Spacks, *Boredom: The Literary History of a State of Mind* (Chicago: University of Chicago Press, 1995).
2. D. D. McNicholl, "Cool's Out for Some Time," *The Australian*, February 4, 2009.
3. Spacks, *Boredom*.
4. Ibid., p. 10.
5. Ibid.
6. Cited ibid., p. 21.
7. Hilary Osborne, "Parental Guidance: Running Cost of a Teenager Is Now £9,000 a Year," *The Guardian*, March 2, 2009.
8. Tamsyn Burgmann, "Teens Willing to Risk Personal Safety to Defend iPods from Muggers," *The Canadian Press*, November 17, 2008.
9. Michelle Higgins and D. J. Booths, "Wii and Other Cool Stuff," *The New York Times*, May 25, 2008.
10. Ibid.
11. Ibid.

CHAPTER FOUR

My iPhone/Myself: Notes from a Digital Fugitive

1. Quoted in Adam Bryant, "He Was Promotable After All," *New York Times*, May 3, 2009.
2. Maria Puente, "The Popularity of Twitter Has Some Relationships in a Twist," *USA Today*, April 17, 2009.
3. Tony Norman, "Man Versus Machine: Who's Winning?" *Pittsburgh Post-Gazette*, March 10, 2009.
4. Galen Gruman, "Can Your iPhone Replace Your Laptop?" InfoWorld, October 5, 2009.
5. Nick Thompson, "Use of Tech Gadgets May Become an Addiction," *Wired*, November 15, 2008.
6. Ibid.
7. Gary Mazo, Martin Trautschold, and Kevin Michaluk, *CrackBerry: True Tales of BlackBerry Use and Abuse* (Charleston, SC: BookSurge, 2008).
8. Ibid.
9. "AOL E-mail Survey: More Than Half of Users Admit They're 'Hooked' on E-mail," *Wireless News*, August 2, 2008.
10. Ibid.
11. Ibid.

12. www.smh.com.au/technology/enterprise/astronauts-tweet-from-space-20100125-mt9k.html.

13. *Wireless News*, 2008.

14. gigaom.com/2010/03/26/1-in-2-americans-will-have-a-smartphone-by-christmas-2011/.

15. Soren Gordhamer, *Wisdom 2.0: Ancient Secrets for the Creative and Constantly Connected* (New York: HarperOne, 2009).

16. "Four in Ten Young Adults Are Mobile-Phone Addicts," *Science News* (University of Granada), February 25, 2007.

17. Kate Stone Lombardi, "Parents' Rights (and Wrongs)," *The New York Times*, July 30, 2006.

18. Bill Marsano, letter to the editor, *The New York Times*, July 20, 2006.

19. Puente, "The Popularity of Twitter Has Some Relationships in a Twist."

20. *News & Notes*, National Public Radio, August 14, 2008.

21. Mary Schmich, "Cell Phone Users: The New Drunk Drivers," *Chicago Tribune*, April 3, 2009.

22. John Naish, *Enough: Breaking Free from the World of Excess* (London: Hodder & Stoughton, 2009).

23. James Harkin, "Tweeting Is Changing the Way We Think," *The Times* (UK), February 18, 2009.

24. James Harkin, *Cyburbia: The Dangerous Idea That's Changing How We Live and Who We Are* (London: Little, Brown, 2009).

25. docs.yahoo.com/docs/pr/release1183.html.

26. Christopher Muther, "Dial 'S' For Shame! Embarrassed by Our Clunky Old Cell Phone, the One That Doesn't Do Anything But (Gasp!) Make Phone Calls? You're Not Alone," *The Boston Globe*, November 13, 2008.

CHAPTER FIVE

The Sound of One Hand Doing Homework

1. Don Tapscott, *Grown Up Digital: How the Net Generation Is Changing Your World* (New York: McGraw-Hill, 2008).

2. Quoted by Laura Bickle in "The Cyber Family," *Today's Parent* (Toronto), 26, no. 1 (June 2009).

3. Cited in Gary Small and Gigi Vorgan, *iBrain: Surviving the Technological Alteration of the Modern Mind* (New York: HarperCollins, 2008), p. 120.

4. Ibid.

5. Cited in Nicholas Carr, "Is Google Making Us Stupid? What the Internet Is Doing to Our Brains," *The Atlantic*, July/August 2008.

6. "Memory and Books, 1477," Medieval Sourcebook: Accounts of Medieval Literacy and Education, c. 1090–1530. www.fordham.edu/halsall/source/medieval-memory.html.

7. Cited in Maggie Jackson, *Distracted: The Erosion of Attention and the Coming Dark Age* (New York: Prometheus Books, 2009).

8. Malcolm Gladwell, "Brain Candy: Is Pop Culture Dumbing Us Down or Smartening Us Up?" *The New Yorker*, May 16, 2005.

9. Roberts, Foehr, and Rideout, "Generation M."

10. Small and Vorgan, *iBrain*.

11. Quoted in Claudia Wallis, "The Multitasking Generation," *Time*, March 19, 2006.

12. Eyal Ophir, Clifford Nass, and Anthony D. Wagner, "Cognitive Control in Media Multitaskers," *Proceedings of the National Academy of Science*, August 2009.

13. Quoted by Ruth Pennebaker, "The Mediocre Multitasker," *The New York Times*, August 30, 2009.

14. Ibid.

15. Lee Shumow, Jennifer A. Schmidt, and Hayal Kackar, "Adolescents' Experience Doing Homework: Associations Among Context, Quality of Experience, and Outcomes," *The School Community Journal*, 18, no. 2 (Fall/Winter 2008).

16. Cited by Catherine Woulfe, "Schools Slash After-Class Work," *Sunday Star Times* (New Zealand), March 3, 2009.

17. Ibid.

18. Rick Docksai, "Teens and Cell Phones," *The Futurist* (Washington), 43, no. 1 (January/February 2009).

19. Carr, "Is Google Making Us Stupid?"

20. Ibid.

21. Jackson, *Distracted*.

22. Ibid.

23. M. Asselin and M. Moayeri, "Toward a Pedagogy for Using the Internet to Learn: An Examination of Adolescent Internet Literacies and Teachers, Parents, and Students' Recommendations for Educational Change," *Proceedings of the Annual Conference of the International Association of School Librarianship*, 2008.

24. Shumow, Schmidt, and Kackar, "Adolescents' Experience Doing Homework."

25. Docksai, "Teens and Cell Phones."

26. Michael Osit, *Generation Text: Raising Well-Adjusted Kids in an Age of Instant Everything* (New York: AMACOM, 2008).

27. Ibid.

28. Ibid.

29. Tamar Lewin, "Study Finds Teenagers' Internet Socializing Isn't Such a Bad Thing," *The New York Times*, November 20, 2008.

30. Ibid.

31. Small and Vorgan, *iBrain*.

32. Selene M. Finch, "A Qualitative Phenomenological Analysis of Modern Communication: Instant Messaging's Importance for Adolescent and Young Adults," doctoral dissertation, University of Phoenix, September 2008.

33. Docksai, "Teens and Cell Phones."

34. Ibid.

35. Wendy K. Kleinman, "Reading, Writing and iPods? How School Erases Boundaries," McClatchy-Tribune Business News, February 18, 2008.

36. Ibid.

37. Wallis, "The Multitasking Generation."

38. Mark Bauerlein, *The Dumbest Generation: How the Digital Age Stupefies Young Americans and Jeopardizes Our Future (Or, Don't Trust Anyone Under 30)* (New York: Tarcher, 2008).

39. Ibid.

40. Ibid.

CHAPTER SIX

Loss of Facebook: Friending the Old-fashioned Way

1. Quoted in James Harkin, *Cyburbia: The Dangerous Idea That's Changing How We Live and Who We Are* (London: Little, Brown, 2009).

2. Quoted in Carr, "Is Google Making Us Stupid?"

3. Small and Vorgan, *iBrain*.

4. Oscar Ybarra, Eugene Burnstein, Piotr Winkielman, Matthew C. Keller, Melvin Manis, Emily Chan, and Joel Rodriguez, "Mental Exercising Through Simple Socializing: Social Interaction Promotes General Cognitive Functioning," *Personality and Social Psychology Bulletin*, 34, no. 248 (2008).

5. Small and Vorgan, *iBrain*.

6. E. Orr, M. Sisic, C. Ross, M. Simmering, J. Arseneault, and R. Orr, "The Influence of Shyness on the Use of Facebook in an Undergraduate Sample," *CyberPsychology and Behavior*, 12, no. 3 (2009), 337−340.

7. "Social Isolation Leading to Violence/Recent Spate of Random Attacks Carried Out by Introverts of All Ages," *Daily Yomiuri* (Japan), July 11, 2009.

8. Ibid.

9. Small and Vorgan, *iBrain*.

10. Ibid.

11. Rebecca Berg, "Autism—An Environmental Health Issue After All?" *Journal of Environmental Health*, June 1, 2009.

12. Roberts, Foehr, and Rideout, "Generation M."

13. "Girls Trapped in Storm Drain Use Facebook to Call for Help . . . Instead of Phoning Emergency Services," *Daily Mail Reporter* (UK), September 8, 2009.
14. "Social Networks Anonymous," *The Economist*, February 2009.
15. Bessie Recep, "You Were Cuter on Facebook," *Cleo*, October 2008.
16. "Primates on Facebook," *The Economist*, February 2009.
17. Neil Seeman, "Facebook and Friendship," *National Post* (Toronto), September 15, 2009.
18. www.insidefacebook.com.
19. Clive Thompson, "Brave New World of Digital Intimacy," *The New York Times*, September 7, 2008.
20. Ferris Jabr, "The New Rules of Social Networking," *Psychology Today*, 41, no. 6 (November/December 2008).
21. Mara E. Zazzali-Hogan and Jennifer Marino Thibodaux, "Ethical Issues to Consider When 'Friending' Witnesses Online," *New Jersey Law Journal*, August 31, 2009.
22. Laura Saunders, "Is 'Friending' in Your Future? Better Pay Your Taxes First," *The Wall Street Journal*, August 27, 2009.
23. Cited in Recep, "You Were Cuter on Facebook."
24. www.cyberbullying.us/research.php.
25. Australian Communications and Media Authority, "Click and Connect: Young Australians' Use of Online Social Media," July 2009.
26. Horatia Harrod, "The World's Photo Album," *Sunday Telegraph* magazine (UK), March 22, 2009.
27. Thompson, "Brave New World."
28. Ibid.
29. Ibid.
30. Ray Kurzweil, *The Singularity Is Near: When Humans Transcend Biology* (New York: Viking, 2006).

CHAPTER SEVEN

Eat, Play, Sleep

1. "Australian Family Dinners End in Arguments," *Herald Sun* (Australia), July 2, 2009.
2. Shira Feldman, Marla E. Eisenberg, Dianne Neumark-Sztainer, and Mary Story, "Associations Between Watching TV During Family Meals and Dietary Intake Among Adolescents," *Journal of Nutrition Education and Behavior*, 39, no. 5 (September/October 2007).
3. Kennedy et al., "Networked Families."
4. "Australian Family Dinners End in Arguments."

5. Feldman et al., "Associations Between Watching TV."

6. Eileen FitzPatrick, Lynn S. Edmonds, and Barbara A. Dennison, "Positive Effects of Family Dinner Are Undone by Television Viewing," *Journal of the American Dietetic Association*, 107, no. 4 (April 2007).

7. David P. Chitakunye and Pauline Maclaran, "The Everyday Practices Surrounding Young People's Food Consumption," *Young Consumers*, 9, no. 3 (2008).

8. Ibid.

9. "University of Minnesota Research Finds Teens Who Have TV in Their Bedroom Are Less Likely to Engage in Healthy Habits," *NewsRx Science*, April 21, 2008.

10. Chitakunye and Maclaran, "The Everyday Practices Surrounding Young People's Food Consumption."

11. Steven Eggermont and Jan Van den Bulck, "Nodding Off or Switching Off? The Use of Popular Media as a Sleep Aid in Secondary-School Children," *Journal of Pediatrics and Child Health*, 42, no. 7 (July 2006).

12. Christina J. Calamaro, Thornton B. A. Mason, and Sarah J. Ratcliffe, "Adolescents Living the 24/7 Lifestyle: Effects of Caffeine and Technology on Sleep Duration and Daytime Functioning," *Pediatrics*, 123 (2009).

13. Raija-Leena Punamäki, Marjut Wallenius, Clase-Håkan Nygård, Lea Saarni, and Arja Rimpelä, "Use of Information and Communication Technology (ICT) and Perceived Health in Adolescence: The Role of Sleeping Habits and Waking-Time Tiredness," *Journal of Adolescence*, 30, no. 4 (August 2007).

14. Robert E. Roberts, Catherine Ramsay Roberts, and Hao T. Duong, "Sleepless in Adolescence: Prospective Data on Sleep Deprivation, Health and Functioning," *Journal of Adolescence*, 32, no. 5 (2009), 1045–1057.

15. Ibid.

16. Calamaro et al., "Adolescents Living the 24/7 Lifestyle."

17. Ibid.

18. Ibid.

CHAPTER EIGHT

The Return of the Digital Native

1. Natasha Bita, "Ban Television for Toddlers," *The Australian*, October 12, 2009.

2. Spacks, *Boredom*.

» Recommended Reading

Australian Communications and Media Authority. "Click and Connect: Young Australians' Use of Online Social Media." July 2009.

Bauerlein, Mark. *The Dumbest Generation: How the Digital Age Stupefies Young Americans and Jeopardizes Our Future (Or, Don't Trust Anyone Under 30)*. New York: Tarcher, 2008.

Carr, Nicholas. "Is Google Making Us Stupid? What the Internet Is Doing to Our Brains." *The Atlantic*, July/August 2008.

———. *The Shallows: What the Internet Is Doing to Our Brains*. New York: W.W. Norton, 2010.

Conley, Dalton. *Elsewhere, USA*. New York: Pantheon, 2009.

Gordhamer, Soren. *Wisdom 2.0: Ancient Secrets for the Creative and Constantly Connected*. New York: HarperOne, 2009.

Harkin, James. *Cyburbia: The Dangerous Idea That's Changing How We Live and Who We Are*. London: Little, Brown, 2009.

Jackson, Maggie. *Distracted: The Erosion of Attention and the Coming Dark Age*. Amherst, NY: Prometheus, 2009.

Johnson, Steven. *Everything Bad Is Good for You*. New York: Riverhead Books, 2006.

Jones, Sydney, and Susannah Fox. "Generations Online in 2009." Pew Internet & American Life Project. January 28, 2009.

Lenhart, Amanda. "Teens and Mobile Phones over the Past Five Years: Pew Internet Looks Back." Pew Internet & American Life Project. August 19, 2009.

Macgill, Alexandra Rankin. "Parents, Teens and Technology." Pew Internet & American Life Project. October 24, 2008.

Mazo, Gary, Martin Trautschold, and Kevin Michaluk. *CrackBerry: True Tales of BlackBerry Use and Abuse*. Charleston, SC: BookSurge, 2008.

McLuhan, Marshall. *Understanding Media: The Extensions of Man*. New York: New American Library, 1964.

Naish, John. *Enough: Breaking Free from the World of Excess*. London: Hodder & Stoughton, 2009.

Ophir, Eyal, Clifford Nass, and Anthony D. Wagner. "Cognitive Control in Media Multitaskers." *Proceedings of the National Academy of Science*, August 2009.

Osit, Michael. *Generation Text: Raising Well-Adjusted Kids in an Age of Instant Everything*. New York: AMACOM, 2008.

Palfrey, John, and Urs Glasser. *Born Digital: Understanding the First Generation of Digital Natives.* New York: Basic Books, 2008.

Postman, Neil. *Amusing Ourselves to Death: Public Discourse in the Age of Show Business.* New York: Elisabeth Sifton Books/Viking, 1985.

————. *The Disappearance of Childhood.* New York: Vintage/Random House, 1994.

Roberts, Donald F., Ulla G. Foehr, and Victoria Rideout. "Generation M: Media in the Lives of 8–18 Year-olds." The Henry J. Kaiser Family Foundation. March 2005.

Small, Gary, and Gigi Vorgan. *iBrain: Surviving the Technological Alteration of the Modern Mind.* New York: HarperCollins, 2008.

Spacks, Patricia Meyer. *Boredom: The Literary History of a State of Mind.* Chicago: University of Chicago Press, 1995.

Tapscott, Don. *Grown Up Digital: How the Net Generation Is Changing Your World.* New York: McGraw-Hill, 2008.

Thoreau, Henry David. *Walden.* Edited by Sherman Paul. Cambridge, MA: The Riverside Press, 1957. (First published in 1854.)